Digital Entertainment

Subhankar Das · Saikat Gochhait
Editors

Digital Entertainment

The Next Evolution in Service Sector

palgrave
macmillan

Editors
Subhankar Das
The Honors Programme
Duy Tan University
Da Nang, Vietnam

Saikat Gochhait
Symbiosis Institute of Digital and
Telecom Management
Symbiosis International University
Pune, Maharashtra, India

ISBN 978-981-15-9723-7   ISBN 978-981-15-9724-4 (eBook)
https://doi.org/10.1007/978-981-15-9724-4

This Palgrave Macmillan imprint is published by the registered company Springer Nature Singapore Pte Ltd.
The registered company address is: 152 Beach Road, #21-01/04 Gateway East, Singapore 189721, Singapore

Preface

Digital entertainment (DE) includes Internet-based gaming, wireless gaming, online discussion clubs for sports or music fans, and any other form of consumer-to-consumer (C2C) entertainment that involves human–computer or human–human interaction via the Internet (or wireless). According to a recent article in Financial Times, the corporate spending on IT has become stagnant in recent years while the market for consumer technologies maintains a strong growth trend. DE is an especially bright spot among the fastest growing business models targeting the consumer market. For instance, five years into the U.S. market, Sony's popular online game, EverQuest, has already attracted 400,000 subscribers and is expected to earn Sony up to $500 million in eight years. This stellar growth is not without problems: many early business developments on DE, even the ones from the largest and most experienced game developers such as Electronic Arts, have faltered. These failures can cost DE companies tens of millions of dollars. Past failures and successes seem to suggest that the success of DE depends not only on solid DE systems development that deliver competitive technological performance and enduring entertainment content, but also on deep comprehension of DE systems adoption and usage by consumers who ultimately decide the fate of any DE product. Newer generations of DE systems, such as World of Warcrafts from Blizzard Entertainment, Netflix, Amazon Prime,

Voot, Zee5, Alexa, Google home have greatly improved their technological performance by tapping into the latest computing and communication technologies. Nevertheless, huge gaps exist in our understanding of how to make DE systems and content more entertaining while controlling the development cost. Furthermore, DE providers often fail to capitalize on their investment. For instance, Internet-based board games, while popular, have contributed little profit to vendors as consumers generally shun fee-based games. Finally, there is a lack of understanding about the roles of DE communities in DE markets. After all, DE users typically interact with a community of peers, a feature that distinguish DEs from stand-alone entertainment or TV-based entertainment. With Covid-19, the primary entertainment source for all is now these over the top (OTT) platforms who stream content directly to digitalized consumers via Internet bypassing all cable television process and satellite entertainment sources.

Next, we center around entries inspecting vivid digital excitement, starting with computer games. Computer games can be considered the quintessential advanced stimulation medium. The articles we incorporated into this territory inspect how these virtual encounters influence us mentally both inside and outside of the game world. We close with three articles on vivid 3-D virtual conditions. These articles inspect mental procedures and impacts of virtual universes as spoke to in vivid, three-dimensional, interceded conditions. The articles in this topical gathering center around the mental procedures and impacts of diversion in VR, just as contrasting VR encounters with nondigital, nonmedicated encounters. We trust that, altogether, this unique book features new edge analysis in advanced excitement in a manner that shows what we know are as yet finding out about stimulation in the digital world. Then we will look for new integrations with Deep learning, Big Data, AI & ML research aspect for digital entertainment. How they will affect the digital entertainment more in future.

Da Nang, Vietnam — Subhankar Das
Pune, India — Saikat Gochhait

Learning Objective of the Book

The purpose of this book is to bring together industry experts and IS researchers to

[1] Introduce the development of the DE industry to an IS audience and discuss the problems encountered in this development process.
[2] Lay out an array of new research venues around DE systems and communities
[3] Discuss the impact of DE on individuals and the society.

We will argue that DE represents an underexplored, new territory and IS researchers are well-positioned to tackle the aforementioned problems with our interdisciplinary background and accumulated knowledge in technology systems.

Highlights of the Book

[1] No book is available on this topic as a whole.
[2] Content and scope will give maximum awareness to various aspects of digital entertainment.
[3] It will solve the basic and advance requirements of digital entertainment in Interactive medium as well as future research in all.
[4] The USP lies in the overall content and structure of book.

Contents

Notes on Contributors

Muhammad Adeel has been working in National Bank of Pakistan since 2016 currently designated as Relationship Manager (C&SME). He has working experience of more than 10 years in the field of IT & General Administration and holding Microsoft Certifications (MCSE, MCITP). He has general interest and research in Wireless Communications, Networking, HCI areas.

Shama Andleeb working as instructor computer in Technical Education and Vocational Training Authority (TEVTA) since August 2011.She has over nine years' experience as trainer, National certified CBT&A assessor. She is currently performing the duties as Head of Institute. She has general interest and research in IoT, blockchain, computer architecture areas.

Prof. Avinash Aslekar is Associate Professor at with Symbiosis Institute of Digital & Telecom Management, Symbiosis International Deemed University, Pune. His areas of interest are Networking, Information Technology, Cybersecurity, and Business intelligence. The author is a Software Developer and Implementer conducting corporate training in the area of Networking and Cybersecurity.

Shariq Aziz Butt Received the Master's Degree from Superior University, Lahore, Pakistan, in 2016, and Ph.D. degree from The University of Lahore, Pakistan. Research collaborations with the Pakistan Institute of Engineering and Applied Sciences, Derby University, UK, Covenant

University, Nigeria and the Universidad De La Costa, CUC, Columbia with grant.

Dr. Subhankar Das is an Academician, Researcher, Author, Writer, Blogger, Data Science trainer, and Social Media Marketing Consultant. He is currently working as Professor and Researcher in the Honors Program, Duy Tan University, Da Nang, Vietnam, with 14+ Years of Teaching and Research Experience and more than 26 Scopus SSCI and SCIe publications.

Dr. Tripti Prabhakar Dhote is an Assistant Professor in the Symbiosis Institute of Digital & Telecom management, Symbiosis International (Deemed University), Pune, India. She has published research papers and case studies in indexed journals. She heads the Entrepreneurship Cell. She is empanelled on the MTDC (Govt of Maharashtra) Tourism Branding committee.

Öğr. Üyesi Orhan Duman is working as Lecturer in Seyfettin Faculty of Applied Sciences Media and Communication Department, Bandırma Onyedi Eylül University Ömer.

Dr. Saikat Gochhait teaches at Symbiosis Institute of Digital & Telecom Management, Symbiosis International Deemed University Pune, India. He is a Ph.D and Post-Doctoral Fellow from the UEx, Spain. He was Awarded DITA and MOFA Fellowship in 2017 and 2018. His research publication with foreign authors indexed in Scopus, ABDC, and Web of Science. He is an IEEE member.

He has more than 50 publications in journals indexed in Scopus, ABDC, Google Scholar, etc. And 6 books published with Springer and IGI Global indexed in Scopus.

Prof. Giri Gandu Hallur works as the Deputy Director and Associate Professor with Symbiosis Institute of Digital & Telecom Management, Symbiosis International (Deemed University), Pune. Ph.D. in Telecom Policy & Regulation. 19 years of work experience as an academician and published more than 25 papers in international conferences and journals.

Dr. Sachin Kumar is currently working as Assistant Professor in the Department of Management Studies, National Institute of Technology Hamirpur, Himachal Pradesh, India. His major area of Interest is Green Marketing, Digital Marketing, Online education, Organic Food and Technology Adoption.

Neha Reddy Kuthuru is a Student in School of Management and Entrepreneurship, Shiv Nadar University, Greater Noida, India.

Dr. Teena Mishra is working as Assistant Professor, PSS Central Vocational Institute, Bhopal. She has published many articles in reputed journals. Her interests lie in IMC, brand communication & digital media.

Dr. Subhra R. Mondal (Academician, Researcher, Author, Writer) is a Researcher in the Honors Program at Duy Tan University. She received her Ph.D. from faculty of management, IBCS, SOA University. Her research interests are concerned with brand management, brand equity with innovative tools, smart tourism, destination marketing, social media marketing, AI & Web 5.0 & digital recruitment.

Dr. Sangeeta Pandit is a Chartered Accountant by profession. She also works as HOD Finance, Sydenham Institute of Management Studies, Research and Entrepreneurship Education (SIMSREE)B Road, Churchgate, Mumbai-400020. She has published many articles in reputed journals.

Dr. Sandeep Prabhu works as Associate Professor at Symbiosis Institute of Digital and Telecom Management, Pune. His research areas are Service Quality, Customer Experience, and Consumer Behavior. He has completed his Ph.D. from Pune University, Pune in Service Quality Management. He has expertise in data science with specific interest in Neural Networks and Deep Learning.

Dr. S. Ravi is working as Professor in Faculty of Communication, Saint Leo University, Savannah. He has published 12 SSCI publications & has worked in Central Asia, Europe, & USA for more than 20 years in the field of Mass Communication & Digital Media.

Dr. Bikramjit Rishi is working as Associate Professor in Institute of Management Technology (IMT), Ghaziabad, India. He has many good publications to his credit. His core interest lies in digital media & marketing.

Kumar Shalender, Post Doctorate, Ph.D. faculty of Strategic Marketing & Change Management at Chitkara University, Punjab. He is engaged in exploring intricacies of Design Thinking, Innovation Cultures, and Leadership Change. He holds command on the subtle nuances of managing team dynamics & group wisdom for achieving effective business outcomes.

Dr. Ramachandran Siri is working as a Professor in University of New England, New South Wales, Australia and Visiting Professor in North Umbria along with two universities of USA. He is having more than 27 years of teaching & research experience in digital learning, marketing. His interest area is HMI & HCI technologies in digital entertainment.

Dr. Ceren Yegen is working as Associate Professor in Faculty of Communication, Department of Journalism, Mersin University, Turkey. She has more than 25 publications in Scopus, ISI indexed journals. She has interest in text analysis, digital media.

List of Figures

Chapter 6

List of Tables

Chapter 9

Chapter 11

CHAPTER 1

Consumer Engagement in Digital Entertainment: A Systematic Review

Sachin Kumar

1 Introduction

Over past few years the researchers have become passionate in describing the consumer engagement in different contexts. Entertainment and engagement are essential to get/retain audience on mobile devices, called "entergagement" (Mazri September 26, 2016). According to a report of Times Internet-DMAasia published in Economic Times, about 45% of the marketers are putting their utmost efforts to analyze the concept of Customer Engagement (CE) (Nadanyiova & Das, 2020). The FDI in Information and Broadcasting, India has been recorded to 8.71 Billion US Dollars for April 2000–December 2019. The Netflix has attained a growth of 700% in 2018–2019 (IBEF, 2020). Therefore, consumer engagement is a new mode of marketing, particularly on brand engagement in a digital age (Roy November 25, 2019).

S. Kumar (✉)
Department of Management Studies, National Institute of Technology, Hamirpur, India
e-mail: sachin@nith.ac.in

S. Das and S. Gochhait (eds.), *Digital Entertainment*,
https://doi.org/10.1007/978-981-15-9724-4_1

Earlier marketing studies were concentrated around consumer and buying behavior, but the engagement has emerged as a new view of consumer marketing (Gambetti & Graffigna, 2010). The term customer engagement is an unexplored concept (Brodie, Ilic, Juric, & Hollebeek, 2013). Existing studies on CE has examined the consumer's engagement on brands (Brodie et al., 2013; Dessart, Veloutsou, & Morgan-Thomas, 2016; Islam, Rahman, & Hollebeek, 2018; Leckie, Nyadzayo, & Johnson, 2016; Lee, Hosanagar, & Nair, 2018; Leventhal, Wallace, Buil, & de Chernatony, 2014; Martínez-López, Anaya-Sánchez, Molinillo, Aguilar-Illescas, & Esteban-Millat, 2017; Schamari & Schaefers, 2015; Tsai & Men, 2013) social media (Barger, Peltier, & Schultz, 2016; Lee et al., 2018; Liu, Lee, Liu, & Chen, 2018; Tafesse 2016) brands on social media (Tsai & Men, 2017) health (Hurley, Keenan, Martsolf, Maeng, & Scanlon, 2009; Mittler, Martsolf, Telenko, & Scanlon, 2013; Ricciardi, Mostashari, Murphy, Daniel, & Siminerio, 2013) personality (Islam, Rahman, & Hollebeek, 2017) e-commerce (Hu & Chaudhary, 2020) and digital entertainment in game (Boyle, Connolly, Hainey, & Boyle, 2012) computing (Lyu, King, Wong, Yau, & Chan, 2005) MP3/DVD (Ponce, 1999), etc.

The concept of CE has drawn the attention of a massive range of academicians and researchers to do an extensive research work on this topic (Dessart et al., 2016) particularly, in context of online system (Brodie et al., 2013). Different customer-related activities or the factors affecting customer engagement have been identified by Gambetti and Graffigna (2010). Furthermore, consumer engagement research is an advancement of its kind in brand engagement research works (Vivek, Beatty, & Morgan, 2012; Dessart et al., 2016). In spite of this popularity of digital entertainment and consumer engagement concepts, there is very less systematic review available. Thus, the researcher was inquisitive about the literature available on Consumer Engagement in Digital Entertainment (CEDE). The research study purposes to make available a systematic literature review of published literature on CEDE and to present the research trend and scenario in this intact field.

The objectives of the study are:

- Producing the systematic literature on Customer engagement in digital entertainment (CEDE).
- To organize all research publications in an efficient way to make search an easy job for future researchers.

- Classification and representation of research articles on CEDE according to research methodology, citation, publisher, country, year of publication.

1.1 Concept of Consumer Engagement (CE)

Consumer engagement is a new trend of managing a customer (Singh, Mondal, Singh, Sahoo, & Das, 2020; Verhoef, Reinartz, & Krafft, 2010). It consists of five types of factors, namely; product, brand, consumer, material and social media (Barber et al., 2016; Singh, Mondal & Das, 2020). Gambetti and Graffigna (2010) revealed in their study that consumer engagement is more apprehensive about engaging people to a brand or advertising. Additionally, a report by Fipp news explains that CE concept acts as a driver to increase company sales (Brodie et al., 2013; Sharma & Das, 2020). Thus, it is a strategy adopted by companies that takes into account the involvement of customers toward them (Bleier, De Keyser, & Verleye, 2018; Das, 2020). Specifically, it is a many-item idea formed of social, cognitive, behavioral and emotional aspects of consumers (Islam & Rahman, 2016; Mondal, 2020).

1.2 Concept of Digital Entertainment

The perceived value and perceived simplicity of use in TAM are the two factors those define the consumer engagement in entertainment (Boyle et al., 2012; Mondal & Sahoo, 2020). The entertainment media are a mode of enjoyment to the users (Das & Nayyar, 2020; Mekler, Bopp, Tuch, & Opwis, 2014). But the enjoyment can be delivered uninterrupted only if the digital entertainment is protected under copyright (Das, Nayyar, & Singh, 2019; Eskicioglu, Town, & Delp, 2003). Video games are a source of digital entertainment (Natkin, 2006) but nowadays only the virtual reality based applications and games are significantly contributing in digital entertainment (Das & Nayyar, 2019; Valente, Feijó, Ribeiro, & Clua, 2018). The reason may be the computer-centered services for home digital entertainment is enjoyed more (Palazzi, Ferretti, Roccetti, Pau, & Gerla, 2006; Singh, Nayyar, Le, & Das, 2019). Further adding such entertainment technologies play a vital and significant job in progress of computer hardware and software (Macedonia, 2000; Mohanty, Dash, Dash, & Das, 2019). To deliver digital access to the people two things are importantly needed one is smartphone and second is cheap data services (IANS, 2019). Since, internet has appeared as an imperative tool for human entertainment (Merritt, 2010), thus digital entertainment can be delivered to the customers in the form of mobile media including mobile music and mobile social media, (Aguado & Martinez, 2014; Singh, Nayyar, & Das, 2019) mobile video (Aguado & Martinez, 2014) games (Boyle et al., 2012, Mekler et al., 2014; Singh & Das, 2018), Video games (Jain, Jain, & Das, 2018; Natkin, 2006) Netflix (Burroughs, 2019; Das, Mondal, Sahoo, Nayyar, & Musunuru, 2018; Matrix 2014; Mithas & Lukas, 2010; Mondal, Das et al., 2017; Van et al., 2020; Walker et al., 2017).

2 Methodology and Systematic Literature Review

Systematic Literature Review Method is adopted in this research paper. This method is a systematic methodological approach to show what is and what is not known, thus making it different from literature review (Das & Mondal, 2016Denyer & Tranfield, 2009; Gupta, Jena, Samantaray, & Das, 2019). The articles from research journals were analyzed. The databases like Emerald, Springer, Sage, Elsevier, Wiley online library were searched for related articles. The citations of the articles were taken from Google scholar as taken by Singh and Kumar (2014). PRISMA guidelines (Moher, Altman, Liberati, & Tetzlaff, 2011; Mondal, Mall et al., 2017) were adopted to review articles under the below four rules set by Bandyopadhyay and Ray (2019).

i. Searching and identifying articles
ii. Screening
iii. Eligibility
iv. Inclusion for Review.

i. Searching and identifying articles: By using Systematic review of literature, 5 databases were rigorously searched for Consumer engagement in Digital Entertainment (CEDE). The databases selected were Emerald, Elsevier, Sage, Springer, Wiley Online Library. The keyword Consumer Engagement was searched in Title/Abstract/Keywords to find the articles. The search process resulted in 204 articles.

Table 1 Scope of keywords research

For Consumer ➔ engagement	Consumer engagement Engage Consumers					
For Digital ➔ Entertainment	Social Media (Facebook, Whatsapp, Twitter, etc.	Netflix, Youtube, Video	Games	Online Gambling	Digital media	Streaming Services

ii. In the second phase screening of articles was done on the basis of reading the abstract on consumer engagement and digital entertainment services. Since, Digital entertainment is a wider term, thus other keywords were also looked and searched for in the titles and abstracts for analysis purpose. The scope of keywords searched is presented in the Table 1.

The scope of study has taken social media as a digital mode of entertainment for generation Y (Verčič & Verčič, 2013). Social media team works as a digital front for the users (Gerbaudo, 2017). Briones, Kuch, Liu, and Jin (2011) explained in their study that using social media in this digital age is effective for maintaining public relations. To enhance further, Dehghani, Niaki, Ramezani, and Sali (2016) said that digital entertainment advertising on YouTube influences the consumer attraction and act as an entertainment source to the viewers. Streaming services like Netflix is a digital platform which is also a means of consumer entertainment (Smith & Telang, 2018). Moreover, any type of media in the digital form serves as an online entertainment (Cunningham & Craig, 2016).

iii. The eligibility of 204 screened articles was determined on the basis of reading. After reading complete matter of the research papers, some articles which do not rightly relate to consumer engagement and digital entertainment were excluded from the list.

iv. Inclusion for Review: The following criteria were chosen for inclusion process:

- Paper published in research journals. Only the Journal articles are considered excluding (conference articles, book articles)

- Papers published in English language
- Paper collected between 2000–2019 (First Mobile with a touch screen was launched in 2000, According to Lobo, Kaskaloglu, Kim, & Herbert, 2011 adopted by Zapata, Fernández-Alemán, Idri, & Toval, 2015)
- Paper showing consumer engagement in title/abstract/keywords.
- Paper showing social media, online brand community, mobile, e-commerce, Facebook, Games, WhatsApp, YouTube, Netflix and other online/digital entertainment engagement platforms which are associated with digital entertainment.
- Only the articles which were accessed by researcher.

2.1 Papers Selected for Systematic Review Using the Inclusion Criteria

Initially without applying any delimitation on the articles of consumer engagement, total 204 articles were resulted. Later, different steps of data screening were followed and it resulted in 159 articles. Then, the articles showing consumer engagement with respect to social media (WhatsApp, Facebook, Twitter, WeChat), games, advertising, video, YouTube, Netflix in the article title/abstract were considered for further evaluations. The main focus was given on title and abstract of articles only. Thus, 24 articles qualified for final systematic review. The details of articles selected are presented in the Table 2.

3 Results

Total 24 articles were reviewed by using Systematic Review Analysis with respect to author wise citations, publisher & journal wise number of articles, country wise publication, year wise publication and type of digital media used in Consumer Engagement and Digital Entertainment (CEDE).

3.1 Year Wise Number of Publications

The year wise publication of research papers is presented in Fig. 1. Although the smartphone was launched in year 2000, and studies-related consumer engagement resulted in search option from year 2000, but meanwhile also rejected for not meeting the inclusion criteria. It was quite

Table 2 Papers selected for systematic review using the inclusion criteria

Name of data Base	*Searched in*	*Years taken for Research*	*Date of search*	*Total Number of Articles*	*No. of Research Articles*	*No. of Articles (Books, case, conference, etc.)*	*No. of Research Articles selected*
Emerald	T/A/K	2000–2019	10 May 2020–11 May 2020	21	20	1	10
Elsevier	T/A/K	2000–2019	12 May 2020–13 May 2020	44	39	5	7
Sage	T/A/K	2000–2019	14 May 2020	27	22	5	1
Wiley Online Library	T/A/K	2000–2019	14 May 2020	66	57	9	5
Springer	T/A/K	2000–2019	16 May 2020	46	21	25	1
Total				204	159	45	24

**T/A/K = Title/Abstract/Keywords*

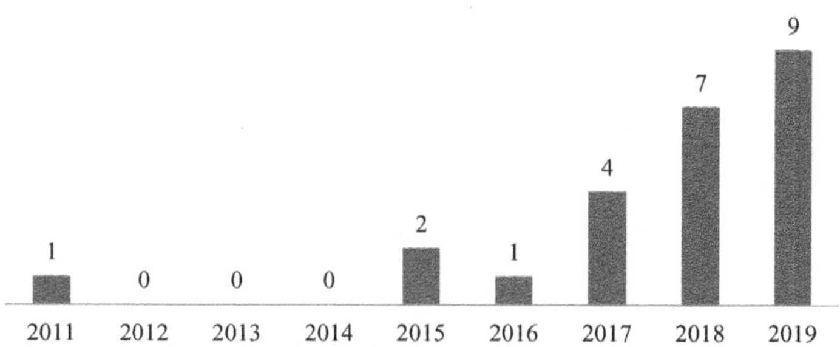

Fig. 1 Year wise number of publications

interesting to see the results that studies related to consumer engagement in digital entertainment were published after 2011. Afterwards, there appeared a dearth of publications on CE for a further period of three years. Later, the publication of articles on the selected subject reappeared and after the year 2017 this number continued to increase till the year 2019. In year 2019, nine papers were published on CEDE and seven in 2018 showing an upward trend.

3.2 Author Wise Citations

The citation means referencing the research work done by other researchers (Singh & Kumar, 2014). The details of researcher wise citation are represented in Table 3. The results illustrate that Ashley and Tuten (2015) from USA score the highest citation 823, followed by Dessart et al. (2015) with 605 citations from UK. It is amazing to note

Table 3 List of author wise citations

Name of Author	*Citation*
Ashley and Tuten (2015)	823
Dessart, Veloutsou, and Morgan-Thomas (2015)	605
Harris and Dennis (2011)	228
Oh, Roumani, Nwankpa, and Hu (2017)	108
Tarute, Nikou, and Gatauti (2017)	75
Tafesse (2016)	67
Islam et al. (2017)	58
Liu et al. (2018)	53
Chen (2017)	35
Phua, Lin, and Lim (2018)	24
Gvili and Levy (2018)	18
Wongkitrungrueng and Assarut (2018)	17
Bianchi and Andrews (2018)	15
Gainsbury, Russell, Hing, and Blaszczynski (2018)	11
Kujur and Singh (2018)	8
Mishra (2019)	8
Moriuchi (2019)	8
Lima, Irigaray, and Lourenco (2019)	7
Liu et al. (2019)	6
Read, Robertson, McQuilken, and Ferdous (2019)	6
Violante, Vezzetti, and Piazzolla (2019)	6
Dessart and Pitardi (2019)	3
Devereux, Grimmer, and Grimmer (2019)	1
Sigurdsson et al. (2019)	1

that Year 2015 reported the highest citations and afterwards the number decreased subsequently till 2019.

3.3 *Publisher and Journal Wise Number of Articles*

The findings in Table 4 indicate that Emerald has published 10 articles on CEDE, followed by 7 articles by Elsevier, 5 by Wiley Online library and 1 each by Sage and Springer, respectively. Springer has published highest number of articles on CEDE but these are not included because of not meeting the inclusion criteria (articles belong to books, conference proceedings). All the selected journals have published one article on CEDE in maximum. But the Journal of Business Research by Elsevier, Journal of Product & Brand Management, Marketing Intelligence & Planning by Emerald, Journal of Consumer Behaviour, Psychology & Marketing by Wiley Online Library have published 2 articles each.

3.4 *Country Wise Publications*

In the Table 5 data represents country wise publications on the basis of countries taken by researchers for the sample collection. It is found that highest numbers of studies (Four) are from USA, followed by Australia, China, UK and India with 3 studies each. The results show that selected articles are from both the developed and developing countries and there is not much difference among the publication frequency of the above mentioned nations.

3.5 *Type of Digital Entertainment Media*

The findings of the articles under study illustrate that the type of digital entertainment media is not restricted to a single category, rather many different types of media platforms are adopted in the selected studies. It is shown in the Fig. 2 that 75% articles are based on Social Media including Facebook, Twitter, Instagram, WeChat, etc., 8% articles were based on YouTube, and 4% each were based on Virtual Reality, Mobile Applications, Google, Online Gambling.

Table 4 Publisher and Journal wise number of articles

Name of Journal	*Elsevier*	*Emerald*	*Sage*	*Springer*	*Wiley Online Library*	*Total*
Computers in Human Behavior	1					1
European Journal of Marketing		1				1
Information & Management	1					1
International Journal of Information Management	1					1
International Journal of Retail & Distribution Management		1				1
International Journal on Interactive Design and Manufacturing				1		1
Internet Research		1				1
Journal of Advances in Management Research		1				1
Journal of Business Research	2					2
Journal of Consumer Behaviour					2	2
Journal of Product & Brand Management		2				2
Managerial and Decision Economics					1	1
Marketing Intelligence & Planning		2				2
New Media & Society			1			1
Online Information Review		1				1
Psychology & Marketing					2	2

(continued)

3.6 Research Methods Adopted

Research methods are the techniques adopted to collect the data for analysis. Figure 3 shows that among the selected 24 papers, 12 papers (50%) adopted questionnaire survey. This survey includes both offline and online

Table 4 (continued)

Name of Journal	*Elsevier*	*Emerald*	*Sage*	*Springer*	*Wiley Online Library*	*Total*
Public Relations Review	1					1
Qualitative Market Research: An International Journal		1				1
Telematics and Informatics	1					1
Grand Total	7	10	1	1	5	24

Table 5 Country wise publications

Country	*Publications*	*Percentage (%)*
USA	4	17
UK	3	13
China	3	13
Australia	3	13
India	3	13
Belgium	1	4
Brazil	1	4
Chile	1	4
Iceland	1	4
Israel	1	4
Italy	1	4
Lithuania	1	4
Thailand	1	4
Grand Total	24	1.00

questionnaire survey i.e., 3 papers (25%) and 9 papers (75%), respectively, in these 12 studies. The second most adopted research method is qualitative analysis with 7 papers (29%) which includes one study (14%) each on focus group, observational research, depth interview, semi-structured interview, respectively, and 3 studies (43%) on content analysis. There are 5 papers (21%) based on secondary data. This secondary data is taken from Social Media Posts (80%) and other different data sources (20%).

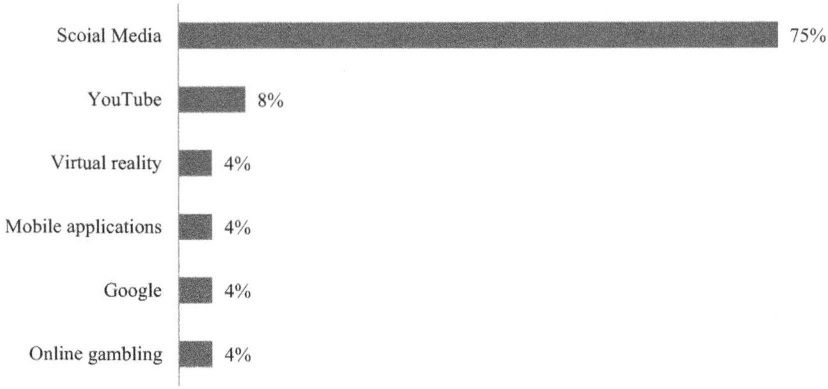

Fig. 2 Type of digital entertainment media

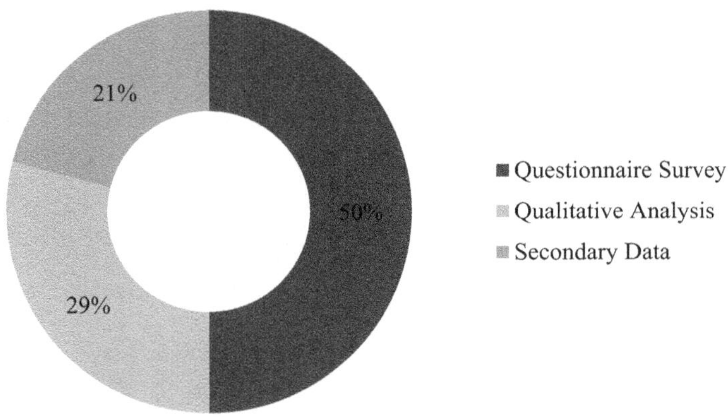

Fig. 3 Type of research method adopted

4 Findings

Consumer engagement is described as the user's stickiness to use the mobile applications Chen (2017). The findings of study reveal that most of the selected research works are from developed countries of the world (Islam & Rahman, 2016), but the count of studies from developing countries like India is quite impressive showing a progressive future for Consumer Engagement in Digital Entertainment. Furthermore, these selected studies did not cover any multi-country for research work, rather focused on a single country only. For Social Media like Facebook, the posts with videos and pictures attract and engage more number of people (Sigurdsson et al., 2019). The engagement is an important element to build brand loyalty among consumers (Liu, Lee, Liu, & Chen, 2019). Besides loyalty, emotional aspect also influences the consumer engagement (Kujur & Singh, 2018). This consumer engagement can be represented by reaction of people to the content, comments on content, content sharing and posting (Barger et al., 2016) because higher level of consumer practical experience with brand create higher level of consumer engagement (Tafesse, 2016). Thus, the firms can take the advantage of videos and emotional appeal to retain higher consumer engagement and taking higher profit margins in India (Das, 2021; Duy et al., 2020).

5 Conclusion

The principal aim of this research was to discover and observe the important features of CEDE and its implications for firms in capturing the consumer market. Systematic Literature Review technique was adopted in this research. Total 204 articles were published on consumer engagement between years 2000–2019 and 24 articles related to title of study were retained for review. It was observed that studies on consumer engagement in digital entertainment started their expansion after year 2015. The citation of the articles selected was highest in 2015 as compared to very fewer citations of latest papers. Maximum numbers of research articles used in this study belong to USA with highest citation, thus presenting USA as a most influential country in this type of research. In 50% of the studies questionnaire survey method has been adopted by the researchers and majority of articles are based on social media. Literature has also revealed that brands can increase the consumer engagement if they use social media. The result has led us to propose some implications to the marketers, more in particular to design the consumer engagement procedures. The videos and emotional appeals can prove as best methods for marketers to attract people toward their brands. As there is a lack of knowledge about human engagement especially in digital entertainment, it can be concluded from this paper that many new studies are necessary to analyze the consumer engagement behavior in digital entertainment.

6 Limitations

Similar to many studies, there are limitations in this study too. First, the present research is based on keyword consumer engagement, but the other similar keywords like customer engagement, public engagement, etc., are not considered because of the title selected. Another limitation is that digital entertainment is represented in most of the previous research works by social media, games, mobile applications, YouTube but not by the online games and online TV channels. The study did not categorize the articles on the basis of type of industry. Moreover only 5 databases were selected and an inclusion criterion of research selection has restricted many research papers to be included.

7 Future Research Scope

This paper provides several insights about consumer engagement. It has presented the detailed framework of year wise expansion of aforesaid topic after 2017. The research has explored that the numbers of publications in this field are increasing in numbers count for last three years, but there are several new areas which can be explored in this field as:

- Research Publications in other databases like Taylor & Francis, IEEE, AMA, Scopus, etc.
- Validity and reliability of the research methods adopted.
- Subject specific review of studies.
- Theories and analysis tools used in the studies.
- Consumer engagement by emotional aspects of product/service usage.
- Empirical research can be conducted on role of digital entertainment on consumer engagement during Lockdown and Covid-19.

References

Aguado, J. M., & Martínez, I. (2014). Feeding digital omnivores: The impact of mobile media in digital entertainment. *The Routledge companion to mobile media* (pp. 181–194). New York and London: Routledge.

Ashley, C., & Tuten, T. (2015). Creative strategies in social media marketing: An exploratory study of branded social content and consumer engagement. *Psychology & Marketing, 32*(1), 15–27.

Bandyopadhyay, C., & Ray, S. (2019). Social enterprise marketing: Review of literature and future research agenda. *Marketing Intelligence & Planning, 38*(1), 121–135.

Barber, N. A., Taylor, D. C., & Remar, D. (2016). Desirability bias and perceived effectiveness influence on willingness-to-pay for pro-environmental wine products. *International Journal of Wine Business Research.*

Barger, V., Peltier, J. W., & Schultz, D. E. (2016). Social media and consumer engagement: A review and research agenda. *Journal of Research in Interactive Marketing, 10*(4), 268–287.

Bianchi, C., & Andrews, L. (2018). Consumer engagement with retail firms through social media: An empirical study in Chile. *International Journal of Retail & Distribution Management, 46*(4), 364–385.

Bleier, A., De Keyser, A., & Verleye, K. (2018). Customer engagement through personalization and customization. *Customer engagement marketing* (pp. 75–94). Cham: Palgrave Macmillan.

Boyle, E. A., Connolly, T. M., Hainey, T., & Boyle, J. M. (2012). Engagement in digital entertainment games: A systematic review. *Computers in Human Behavior, 28*(3), 771–780.

Briones, R. L., Kuch, B., Liu, B. F., & Jin, Y. (2011). Keeping up with the digital age: How the American Red Cross uses social media to build relationships. *Public Relations Review, 37*(1), 37–43.

Brodie, R. J., Ilic, A., Juric, B., & Hollebeek, L. (2013). Consumer engagement in a virtual brand community: An exploratory analysis. *Journal of Business Research, 66*(1), 105–114.

Burroughs, B. (2019). House of Netflix: Streaming media and digital lore. *Popular Communication, 17*(1), 1–17.

Chen, Y. R. R. (2017). Perceived values of branded mobile media, consumer engagement, business-consumer relationship quality and purchase intention: A study of WeChat in China. *Public Relations Review, 43*(5), 945–954.

Cunningham, S., & Craig, D. (2016). Online entertainment: A new wave of media globalization? *International Journal of Communication, 10,* 5409–5425.

Das, S. (2020). Innovations in digital banking service brand equity and millennial consumerism. *Digital transformation and innovative services for business and learning* (pp. 62–79). IGI Global. https://doi.org/10.4018/978-1-7998-5175-2.ch004.

Das, S. (2021). *Search engine optimization and marketing: A recipe for success in digital marketing* (1st ed.). CRC Press. https://doi.org/10.1201/9780429298509.

Das, S., & Mondal, S. (2016). *Yours digitally: Patanjali*. Available at SSRN 2737299.
Das, S., Mondal, S. R., Sahoo, K. K, Nayyar, A. & Musunuru, K. (2018). Study on impact of socioeconomic make up of Facebook users on purchasing behavior. *Revista Espacios, 39*, 28–42. Retrieved from http://www.revistaespacios.com/a18v39n33/18393328.html.
Das, S., & Nayyar, A. (2019). Innovative ideas to manage urban traffic congestion in cognitive cities. In *Driving the development, management, and sustainability of cognitive cities* (pp. 139–162). IGI Global. https://doi.org/10.4018/978-1-5225-8085-0.ch006.
Das, S., & Nayyar, A. (2020). Effect of consumer green behavior perspective on green unwavering across various retail configurations. In *Green marketing as a positive driver toward business sustainability* (pp. 96–124). IGI Global. https://doi.org/10.4018/978-1-5225-9558-8.ch005.
Das, S., Nayyar, A., & Singh, I. (2019). An assessment of forerunners for customer loyalty in the selected financial sector by SEM approach toward their effect on business. *Data Technologies and Applications, 53*(4), 546–561. https://doi.org/10.1108/DTA-04-2019-0059.
Dehghani, M., Niaki, M. K., Ramezani, I., & Sali, R. (2016). Evaluating the influence of YouTube advertising for attraction of young customers. *Computers in Human Behavior, 59*, 165–172.
Denyer, D., & Tranfield, D. (2009). Producing a systematic review. In D Buchanan & A. Bryman (Eds.), *The SAGE handbook of organizational research methods* (pp. 671–689). London: Sage.
Dessart, L., & Pitardi, V. (2019). How stories generate consumer engagement: An exploratory study. *Journal of Business Research, 104*, 183–195.
Dessart, L., Veloutsou, C., & Morgan-Thomas, A. (2015). Consumer engagement in online brand communities: A social media perspective. *Journal of Product & Brand Management, 24*(1), 28–42.
Dessart, L., Veloutsou, C., & Morgan-Thomas, A. (2016). Capturing consumer engagement: Duality, dimensionality and measurement. *Journal of Marketing Management, 32*(5–6), 399–426.
Devereux, E., Grimmer, L., & Grimmer, M. (2019). Consumer engagement on social media: Evidence from small retailers. *Journal of Consumer Behaviour, 19*(2), 151–159.
Duy, N. T., Mondal, S. R., Van, N. T. T., Dzung, P. T., Minh, D. X. H., & Das, S. (2020). A study on the role of web 4.0 and 5.0 in the sustainable tourism ecosystem of Ho Chi Minh City, Vietnam. *Sustainability, 12*(17), 7140. https://doi.org/10.3390/su12177140.
Eskicioglu, A. M., Town, J., & Delp, E. J. (2003). Security of digital entertainment content from creation to consumption. *Signal Processing: Image Communication, 18*(4), 237–262.

Gainsbury, S. M., Russell, A. M., Hing, N., & Blaszczynski, A. (2018). Consumer engagement with and perceptions of offshore online gambling sites. *New Media & Society, 20*(8), 2990–3010.

Gambetti, R. C., & Graffigna, G. (2010). The concept of engagement: A systematic analysis of the ongoing marketing debate. *International Journal of Market Research, 52*(6), 801–826.

Gerbaudo, P. (2017). Social media teams as digital vanguards: The question of leadership in the management of key Facebook and Twitter accounts of Occupy Wall Street, Indignados and UK Uncut. *Information, Communication & Society, 20*(2), 185–202.

Gupta, D. K., Jena, D., Samantaray, A. K., & Das, S. (2019). HRD climate in selected public sector banks in India. *Revista ESPACIOS, 40*(11), 14.

Gvili, Y., & Levy, S. (2018). Consumer engagement with eWOM on social media: The role of social capital. *Online Information Review, 42*(4), 482–505.

Harris, L., & Dennis, C. (2011). Engaging customers on Facebook: Challenges for e-retailers. *Journal of Consumer Behaviour, 10*(6), 338–346.

Hu, M., & Chaudhry, S. S. (2020). Enhancing consumer engagement in e-commerce live streaming via relational bonds. *Internet Research, 30*(3), 1019–1041.

Hurley, R. E., Keenan, P. S., Martsolf, G. R., Maeng, D. D., & Scanlon, D. P. (2009). Early experiences with consumer engagement initiatives to improve chronic care. *Health Affairs, 28*(1), 277–283.

Islam, J. U., & Rahman, Z. (2016). The transpiring journey of customer engagement research in marketing. *Management Decision, 54*(8), 2008–2034.

Islam, J. U., Rahman, Z., & Hollebeek, L. D. (2017). Personality factors as predictors of online consumer engagement: An empirical investigation. *Marketing Intelligence & Planning, 35*(4), 510–528.

Islam, J. U., Rahman, Z., & Hollebeek, L. D. (2018). Consumer engagement in online brand communities: A solicitation of congruity theory. *Internet Research, 28*(1), 23–45.

Jain, S., Jain, V., & Das, S. (2018). Relationship analysis between emotional intelligence and service quality with special evidences from Indian banking sector. *Revista ESPACIOS, 39*(33), 3.

Kujur, F., & Singh, S. (2018). Emotions as predictor for consumer engagement in YouTube advertisement. *Journal of Advances in Management Research, 15*(2), 184–197.

Leckie, C., Nyadzayo, M. W., & Johnson, L. W. (2016). Antecedents of consumer brand engagement and brand loyalty. *Journal of Marketing Management, 32*(5–6), 558–578.

Lee, D., Hosanagar, K., & Nair, H. S. (2018). Advertising content and consumer engagement on social media: Evidence from Facebook. *Management Science, 64*(11), 5105–5131.

Leventhal, R. C., Wallace, E., Buil, I., & de Chernatony, L. (2014). Consumer engagement with self-expressive brands: Brand love and WOM outcomes. *Journal of Product & Brand Management, 23*(1), 33–42.

Lima, V. M., Irigaray, H. A. R., & Lourenco, C. (2019). Consumer engagement on social media: Insights from a virtual brand community. *Qualitative Market Research: An International Journal, 22*(1), 14–32.

Liu, L., Lee, M. K., Liu, R., & Chen, J. (2018). Trust transfer in social media brand communities: The role of consumer engagement. *International Journal of Information Management, 41*, 1–13.

Liu, L., Liu, R., Lee, M., & Chen, J. (2019). When will consumers be ready? A psychological perspective on consumer engagement in social media brand communities. *Internet Research, 28*(1), 23–45.

Lobo, D., Kaskaloglu, K., Kim, C. Y., & Herbert, S. (2011). Web usability guidelines for smartphones: A synergic approach. *International Journal of Information and Electronics Engineering, 1*(1), 33–37.

Lyu, M. R., King, I., Wong, T. T., Yau, E., & Chan, P. W. (2005, March). Arcade: Augmented reality computing arena for digital entertainment. In *2005 IEEE Aerospace Conference* (pp. 1–9). IEEE.

Macedonia, M. (2000). Why digital entertainment drives the need for speed. *Computer, 33*(2), 124–127.

Martínez-López, F. J., Anaya-Sánchez, R., Molinillo, S., Aguilar-Illescas, R., & Esteban-Millat, I. (2017). Consumer engagement in an online brand community. *Electronic Commerce Research and Applications, 23*, 24–37.

Matrix, S. (2014). The Netflix effect: Teens, binge watching, and on-demand digital media trends. *Jeunesse: Young People, Texts, Cultures, 6*(1), 119–138.

Mekler, E. D., Bopp, J. A., Tuch, A. N., & Opwis, K. (2014, April). A systematic review of quantitative studies on the enjoyment of digital entertainment games. In *Proceedings of the SIGCHI Conference on Human Factors in Computing Systems* (pp. 927–936). https://doi.org/10.1145/2556288.2557078.

Merritt, L. R. (2010). *The relationships between media influence, peer pressure, appearance importance, body perception, and clothing behavior: Examing ethnic differences among adolescent females in the united states* (Doctoral dissertation, University of Georgia).

Mishra, A. S. (2019). Antecedents of consumers' engagement with brand-related content on social media. *Marketing Intelligence & Planning, 37*(4), 386–400.

Mithas, S., & Lucas, H. C. (2010). What is your digital business strategy? *IT Professional, 12*(6), 4–6.

Mittler, J. N., Martsolf, G. R., Telenko, S. J., & Scanlon, D. P. (2013). Making sense of "consumer engagement" initiatives to improve health and health care: A conceptual framework to guide policy and practice. *The Milbank Quarterly, 91*(1), 37–77.

Mohanty, P. C., Dash, M., Dash, M., & Das, S. (2019). A study on factors influencing training effectiveness. *Revista Espacios, 40,* 7–15. Retrieved from http://www.revistaespacios.com/a19v40n02/19400207.html.

Moher, D., Altman, D. G., Liberati, A., & Tetzlaff, J. (2011). PRISMA statement. *Epidemiology, 22*(1), 128.

Mondal, S. R. (2020). A systematic study for digital innovation in management education: An integrated approach towards problem-based learning in Vietnam. In *Digital innovations for customer engagement, management, and organizational improvement* (pp. 104–120). IGI Global. https://doi.org/10.4018/978-1-7998-5171-4.ch006.

Mondal, S., Das, S., Musunuru, K., & Dash, M. (2017). Study on the factors affecting customer purchase activity in retail stores by confirmatory factor analysis. *Revista Espacios, 38*(61), 30.

Mondal, S., Mall, M., Mishra, U. S., & Sahoo, K. (2017). Investigating the factors affecting customer purchase activity in retail stores. *Revista Espacios, 38*(57), 22.

Mondal, S., & Sahoo, K. K. (2020). A study of green building prospects on sustainable management decision making. In *Green building management and smart automation* (pp. 220–234). IGI Global. https://doi.org/10.4018/978-1-5225-9754-4.ch011.

Moriuchi, E. (2019). Okay, Google!: An empirical study on voice assistants on consumer engagement and loyalty. *Psychology & Marketing, 36*(5), 489–501.

Nadanyiova, M., & Das, S. (2020). Millennials as a target segment of socially responsible communication within the business strategy. *Littera Scripta, 13*(1), 119–134. https://doi.org/10.36708/Littera_Scripta2020/1/8.

Natkin, S. (2006). *Video games and interactive media: A glimpse at new digital entertainment*. New York: AK Peters/CRC Press.

Oh, C., Roumani, Y., Nwankpa, J. K., & Hu, H. F. (2017). Beyond likes and tweets: Consumer engagement behavior and movie box office in social media. *Information & Management, 54*(1), 25–37.

Palazzi, C. E., Ferretti, S., Roccetti, M., Pau, G., & Gerla, M. (2006). What's in that magic box? The home entertainment center's special protocol potion, revealed. *IEEE Transactions on Consumer Electronics, 52*(4), 1280–1288.

Phua, J., Lin, J. S. E., & Lim, D. J. (2018). Understanding consumer engagement with celebrity-endorsed E-Cigarette advertising on instagram. *Computers in Human Behavior, 84,* 93–102.

Ponce, B. (1999). The impact of MP3 and the future of digital entertainment products. *IEEE Communications Magazine, 37*(9), 68–70.

Read, W., Robertson, N., McQuilken, L., & Ferdous, A. S. (2019). Consumer engagement on Twitter: perceptions of the brand matter. *European Journal of Marketing, 53*(9), 1905–1933.

Ricciardi, L., Mostashari, F., Murphy, J., Daniel, J. G., & Siminerio, E. P. (2013). A national action plan to support consumer engagement via e-health. *Health Affairs, 32*(2), 376–384.

Schamari, J., & Schaefers, T. (2015). Leaving the home turf: How brands can use webcare on consumer-generated platforms to increase positive consumer engagement. *Journal of Interactive Marketing, 30*, 20–33.

Sharma, E., & Das, S. (2020). Measuring impact of Indian ports on environment and effectiveness of remedial measures towards environmental pollution. *International Jouranl of Environment and Waste Management, 25*(3), 356–380. https://doi.org/10.1504/IJEWM.2019.10021787.

Sigurdsson, V., Larsen, N. M., Sigfusdottir, A. D., Fagerstrøm, A., Alemu, M. H., Folwarczny, M., & Foxall, G. (2019). The relationship between the firm's social media strategy and the consumers' engagement behavior in aviation. *Managerial and Decision Economics, 41*(2), 234–249.

Singh, H. P., & Kumar, S. (2014). Working capital management: A literature review and research agenda. *Qualitative Research in Financial Markets, 6*(2), 177–197.

Singh, I., Nayyar, A., & Das, S. (2019). A study of antecedents of customer loyalty in banking & insurance sector and their impact on business performance. *Revista ESPACIOS, 40*(06), 11.

Singh, I., Nayyar, A., Le, D. H., & Das, S. (2019). A conceptual analysis of internet banking users' perceptions. An Indian perceptive. *Revista ESPACIOS, 40*(14), 1.

Singh, L. B., Mondal, S. R., & Das, S. (2020). Human resource practices & their observed significance for Indian SMEs. *Revista ESPACIOS, 41*(07). Retrieved from http://www.revistaespacios.com/a20v41n07/20410715.html.

Singh, S., & Das, S. (2018). Impact of post-merger and acquisition activities on the financial performance of banks: A study of Indian private sector and public sector banks. *Revista Espacios Magazine, 39*(26), 25.

Singh, S., Mondal, S., Singh, L. B., Sahoo, K. K., & Das, S. (2020). An empirical evidence study of consumer perception and socioeconomic profiles for digital stores in Vietnam. *Sustainability, 12*(5), 1716.

Smith, M. D., & Telang, R. (2018). Data can enhance creative projects—Just look at Netflix. *Harvard Business Review.*

Tafesse, W. (2016). An experiential model of consumer engagement in social media. *Journal of Product & Brand Management, 25*(5), 424–434.

Tarute, A., Nikou, S., & Gatautis, R. (2017). Mobile application driven consumer engagement. *Telematics and Informatics, 34*(4), 145–156.

Tsai, W. H. S., & Men, L. R. (2013). Motivations and antecedents of consumer engagement with brand pages on social networking sites. *Journal of Interactive Advertising, 13*(2), 76–87.

Tsai, W. H. S., & Men, L. R. (2017). Consumer engagement with brands on social network sites: A cross-cultural comparison of China and the USA. *Journal of Marketing Communications, 23*(1), 2–21.

Valente, L., Feijó, B., Ribeiro, A., & Clua, E. (2018). Pervasive virtuality in digital entertainment applications and its quality requirements. *Entertainment Computing, 26*, 139–152.

Van, N. T. T., Vrana, V., Duy, N. T., Minh, D. X. H., Dzung, P. T., Mondal, S. R., & Das, S. (2020). The role of human–machine interactive devices for post-COVID-19 innovative sustainable tourism in Ho Chi Minh City, Vietnam. *Sustainability, 12*(22), 9523. https://doi.org/10.3390/su12229523.

Verčič, A. T., & Verčič, D. (2013). Digital natives and social media. *Public Relations Review, 39*(5), 600–602.

Verhoef, P. C., Reinartz, W. J., & Krafft, M. (2010). Customer engagement as a new perspective in customer management. *Journal of Service Research, 13*(3), 247–252.

Violante, M. G., Vezzetti, E., & Piazzolla, P. (2019). How to design a virtual reality experience that impacts the consumer engagement: The case of the virtual supermarket. *International Journal on Interactive Design and Manufacturing (IJIDeM), 13*(1), 243–262.

Vivek, S. D., Beatty, S. E., & Morgan, R. M. (2012). Customer engagement: Exploring customer relationships beyond purchase. *Journal of Marketing Theory and Practice, 20*(2), 122–146.

Walker, R., Jeffery, M., So, L., Sriram, S., Nathanson, J., Ferreira, J., & Feldmeier, J. (2017). Netflix leading with data: The emergence of data-driven video. *Kellogg School of Management Cases, 1*(1), 1–19.

Wongkitrungrueng, A., & Assarut, N. (2018). The role of live streaming in building consumer trust and engagement with social commerce sellers. *Journal of Business Research*. https://doi.org/10.1016/j.jbusres.2018.08.032.

Zapata, B. C., Fernández-Alemán, J. L., Idri, A., & Toval, A. (2015). Empirical studies on usability of mHealth apps: A systematic literature review. *Journal of medical systems, 39*(2), 1.

Online references

ET Online. (August 3, 2018). Customer engagement key focus for 45% marketers in 2018: Times Internet-DMAasia report. *The Economic Times*. Retrieved on May 7, 2020 from https://economictimes.indiatimes.com/industry/media/entertainment/media/customer-engagement-key-focus-for-45-marketers-in-2018-times-internet-dmaasia-report/articleshow/65255111.cms?from=mdr.

IANS. (August 20, 2019). *Indian media, entertainment industry to hit Rs 3 lakh crore by 2024*. Retrieved on May 7, 2020 from https://cio.economictimes.indiatimes.com/news/internet/indian-media-entertainment-industry-to-hit-rs-3-lakh-crore-by-2024/70754728.

IBEF. (March, 2020). *Media and Entertainment industry*. Retrieved on May 6, 2020 from https://www.ibef.org/industry/media-entertainment-india.aspx.

Mazri, N. (September 26, 2016). *Entertainment + Engagement: The key to getting an audience on mobile*. Retrieved on May 6, 2020 from https://www.entrepreneur.com/article/282892.

Roy, D. (November 25, 2019). *Importance of consumer engagement for brand retention in the digital age*. ETBrandEquity. Retrieved on May 6, 2020 from https://brandequity.economictimes.indiatimes.com/news/marketing/importance-of-consumer-engagement-for-brand-retention-in-the-digital-age/72215852.

CHAPTER 2

Review of Consumer Engagement and Digital Entertainment on Over the Top Platforms

Tripti Prabhakar Dhote

1 Introduction

Over the past few decades, Entertainment has undergone a definite transition which is digitally driven. Collinger, Malthouse, and Maslowska (2011) had developed a Consumer Engagement Engine at **the Spiegel research center** which indicates a synergistic collaborative and innovative interaction of consumers in the digital ecosystem influencing customer value. The media and entertainment sector has not only undergone a major transformation in terms of entertainment but is also consistently at the cutting-edge of technological innovation and growth (Nadanyiova & Das, 2020). Digital platforms have emerged as one of the biggest, the most rewarding and profitable, markets for entertainment in the world. (Kuwamura, 2020; Singh, Mondal, Singh, Sahoo, & Das, 2020).

T. P. Dhote (✉)

Symbiosis Institute of Digital and Telecom Management (SIDTM), Symbiosis International (Deemed University), Pune, India

S. Das and S. Gochhait (eds.), *Digital Entertainment*,
https://doi.org/10.1007/978-981-15-9724-4_2

According to a PWC—Outlook report, (2018) Spending on entertainment and media in countries like India has increased manifold fuelled by a growing middle class and a young urban population. However despite this growth, the entertainment and media market in India has been lagging being Asia Pacific markets like China (Singh, Mondal, & Das, 2020). Over the past five years India has shown a huge overall spending on entertainment and media growing at of 11.6% (CAGR) which is almost double the 5.0% CAGR as estimated for global entertainment and media (Sharma & Das, 2020).

With data becoming cheaper and the proliferation of various devices more and more content is being produced globally. The launching of Apple news + App in 2019 provided consumers an easy access with a certain fixed subscription fee to over 300 newspapers, magazines and digital publications. While digital streaming services like Spotify, Netflix, Hotstar are also available across different devices, A PWC Outlook report (2019) claims that in South East Asia in particular, Telecom companies are providing bundled data in an affordable package with Internet OTT services and other forms of entertainment packed together (Das, 2020).

While previously there were limited choices for consumers, but with the gradual evolution and rise of digital platforms, the approaches and formats of entertainment are also undergoing change (Mondal, 2020). Digital technology has made an aggressive foray into the entertainment domain completely disrupting the space. (Katiyal, 2020; Mondal & Sahoo, 2020). With technology converging, overlapping and touching the minutest aspect of consumer lives, the focus of entertainment in the current context appears to be more on human-centric design where technology plays the binding factor. (Roy, 2019; Das & Nayyar, 2020). As per a KPMG report (2019) a sudden splurge has been observed in the digital and media-entertainment platforms, especially the over the top (OTT) audio and video segments. While the rapid access and proliferation of mobile technology coupled with cheaper data rates has fueled on demand content consumption (Das, Nayyar, & Singh, 2019).

This has also brought about considerable observable changes not only in the way consumer demand is addressed but also how digital entertainment is delivered across platforms, giving way for an increased inclination towards creating a virtual ecosystem for consumer engagement and expression (Das & Nayyar, 2019). A near complete digital transformation has also brought about a transition in the consumption journey of the consumer. Content is no longer centered on the television platform and the delivery method too is no longer driven by a single content provider or even a cord to connect to it. (Duy et al., 2020; Hughes, 2019; Singh, Nayyar, Le, & Das, 2019; Van et al., 2020).

One of the possible reasons could be attributed to the consumers having a stronger voice, more control hence becoming more demanding and the resultant inability of traditional above—the line entertainment platforms to hold on to the attention span (Mohanty, Dash, Dash, & Das, 2019). Also with a tremendous choice, user friendly features, seamless accessibility and an ease of consumption even on the go, trends like binge-watching are rapidly gaining momentum. (Katiyal, 2020; Singh, Nayyar & Das, 2019). Therefore while Above the Line (ATL) platforms like Television have always focused on creating content for entertainment aimed at stickiness and consumer retention, evolving Through-the line (TTL) platforms like OTT (Over the Top) and other video platforms like YouTube are largely driven on strengthening customer relationships by way of specific viewer targeting based on their profiles and viewing habits (Roy, 2019; Gupta, Jena, Samantaray, & Das, 2019). Statistics claim that the OTT sector in India alone saw a rise of 59% in 2019, which accounts to Rs 17 billion and is expected to further grow and touch the 24 billion mark by 2021. The exponential growth of OTT technology is a clear indication of the changing media consumption patterns. Over The Top (OTT) platforms have brought about a metamorphosis in the larger media and entertainment domain by offering diversity of content while creating an immersive and engaging experience for consumers by actually transitioning the way content is created distributed and consumed. (Katiyal, 2020; Singh & Das, 2018). The demand for content services inclusive of behind the camera functions like editing, post production, VFX, animation driving facilities like uplink, content analytics, tagging, dubbing, formatting, along with Virtual reality (VR), Augmented Reality (AR) various gaming services and a variety of apps is creating huge engagement opportunities and is expected to clock US$2.5 to US$3 billion by 2025. (Jain, Jain, & Das, 2018; Pherwani, 2020).

Though technology has always facilitated an innovative digital experience in terms of the options and content yet the human experience striking relevant relationship has been missing. There is a remarkable shift in content today which keeps consumers at the core and is focussed on a human centered approach ushering in an incredible revelation in the way consumers are engaged. (Das, Mondal, Sahoo, Nayyar, & Musunuru,

2018; Roy, 2019). Entertainment in the current digital context is definitely moving away from the traditional norms by delivering engaging experiences that are very human, empathy driven and interactive. Social media in particular has evolved as one of the most defining technologies in the current context. With free interactive spaces it offers various possibilities for consumer engagement. (Appel, Grewal, Hadi, & Stephan, 2020). The consumer is more empowered with an influential voice and has powerful diverse options to choose from a wide mix like, price bundles, delivery endpoints and accounts to select the specific entertainment of their choice. (Das & Mondal, 2016; Hughes, 2019).

Entertainment is therefore gradually becoming more and more personalized with an easy accessibility on diverse social sites like Instagram, Twitter and Facebook. Consumer is being served with versatile engagement options (Mondal, Das et al., 2017). While this is viewed as a challenge from the consumer engagement perspective, it is also looked at as an opportunity to develop emotionally intelligent platforms for stronger human connect by designing the right experience (Das, 2021; Mondal, Mall et al., 2017; Roy, 2019).

2 Review of Literature

2.1 Consumer Engagement: The Theoretical Framework

Consumer engagement has its roots in the concept of Engagement, which has been widely studied and observed by researchers in different contexts as the driver to motivation and finds applications in varied domains including organizational behavior for enhanced organizational productivity (Schaufeli, Salanova, Gonza´lez, & Bakker, 2002) therefore realizing the need for an intellectual engagement of employees with their respective jobs (Hardikar & Fill, 2005). Studies also look at engagement as an attribute of energy, involvement and efficiency (Maslach & Leiter, 1997). Schaufeli et al. (2002) in their research also view engagement as an all—encompassing state with a dual character which is both cognitive and affective and does not have a reference to a specific individual an object or a particular behavior. In the digital context consumer engagement has been interpreted as a vital component of Relationship Marketing. (Ashley, Christy, Noble, Donthu, & Lemon, 2011; Vivek, Beatty, & Morgan, 2012).

Literature observes technology as one of the most crucial aspect in consumer engagement. Davis (1989) in his study on engagement in digital entertainment games has proposed a theory that finds utility and perceived use as the important indicators of technology acceptance. Researchers also indicate the multifaceted multidimensional and dynamic traits of consumer engagement, and also reveal that engagement may exist in varied states depending on its degree of intensity over time (Brodie, Ilic, Juric, & Hollebeek, 2013).

With the growing impact of customer engagement on customer value **Spiegel research Center** in their study observe that in today's context of digital communication engagement functions like an engine offering brands and consumers a platform to interact with each other in an innovative manner thus influencing customer value. Given the increased emphasis on the brand's efforts for fostering relationships with consumers by offering tailored interactions through numerous platforms the concept of consumer engagement is gaining a lot of traction (Dessart, Veloutsou, & Thomas, 2016). The ever increasing role of social media and its interactivity with brands has further propelled several different perspectives of engagement with reference to consumers. One of the studies which focuses on the psychological aspect addresses consumer engagement as a bonding and continuous relationship of consumers with brands (Brodie, Ilic, Juric, & Hollebeek, 2011) Braojos-Gomez, Benitez-Amado, and Llorens-Montes (2015) in one of their earlier studies identify two different forms of consumer engagement online. While social engagement as per the study is fuelled by social websites like Facebook and Twitter, Conventional engagement is essentially driven by web-based technologies like a company's website.

The growing influence of the consumer's voice in the digital context has made several existing assumptions related to consumer engagement outdated. From the practioners perspective, the construct of Consumer engagement focuses on leveraging and utilizing digital tools like social media for interacting, socializing and content creation to build a strong relationship with Brands (Armano, 2008). Garreston (2008) advocates that social media is being widely used by consumers not just for researching on products and brands before purchase but also as a platform for engaging with other consumers for reviews and important inputs. Vivek et al. (2012), outline that the value of consumer engagement lies in interactive and experiential consumer relationships.

However, with the evolution of various platforms for entertainment the concept has undergone a transition and it has been addressed beyond an outcome-based concept that includes connect and relationship. Studies also look at Site visits, page views, recommendations, attention ability, frequency, time spend on a digital site as well as the behavioral usage as major indicators of consumer engagement. The suggested definition of engagement with media therefore, is diverting a potential consumer to a brand idea enhanced by the background context.

Literature clearly observes social media as the impetus for engagement and entertainment revolution in the digital media (Vogt & Knapman, 2008). There are numerous studies which relate consumer engagement on social media to customer satisfaction (Challagalla, Venkatesh, & Kohli, 2009) that can bring about a strong word of mouth (Cheung, Lee, & Jin, 2011). Studies also point out that when it comes to drawing consumer attention the concept of engagement is largely impacted by Involvement and interactivity. (Abdul-Ghani, Hyde, & Marshall, 2011). While Van Doorn et al. (2010) suggest that consumer engagement is a motivated behavior that extends beyond purchase and consumption.

The experience that a consumer gets is also counted as a significant measure of the extent of engagement in the digital environment, This subsequently has given rise to a wide spectrum of platforms that are interactive in nature ranging from content communities like You tube to microblogging sites like twitter and social networking sites like Facebook (Avery et al. 2010).

2.2 Understanding Digital Entertainment

Driven by digitisation and an ever increasing content consumption the spend on digital media and entertainment has grown immensely in recent times. Apart from establishing a personal connect with others (Alvarez & Fournier, 2016; Kumar & Gupta 2016; Lamberton & Stephen 2016), studies suggest several behavioral expressions of digital entertainment like a blog a video, likes, post sharing and various different forms of content. (Hollebeek, Glynn, & Brodie, 2014). There is no scientific definition available on the term digital entertainment in literature which is a limitation. However, for the purpose of this chapter the understanding has been derived and compiled from varied sources where digital entertainment draws its base from certain enabling fundamental technologies like

Internet and computers, the World Wide Web, multimedia and streaming media (Aki Järvinen, Satu Heliö, & Frans Mäyrä, 2002).

2.3 *Leveraging Technology for Engagement*

The metamorphic growth in different emerging technologies has tremendously impacted entertainment as also consumer consumption patterns. While Machine Learning, Artificial Intelligence (AI), Virtual Reality (VR), IOT, Augmented Reality (AR) are evolving and making rapid strides into the digital entertainment space, the emergence of other futuristic technology like block chain coupled with 5G will definitely facilitate and push content creation and distribution. Literature emphasizes on the enabling role of convergence of content, gaming and other interactive services to propel real-time consumption of digital media. Several organizations like Infosys have their own in built "experience studio" for mobile applications to facilitate better user engagement. Technology in the current context facilitates seamless engagement across platforms with features which can allow consumers to restore, replace reproduce, recreate and conjure virtual items to deliver immersive experiences through multiple devices such as smart mobiles, interactive screens, game consoles and wearables devices like headsets. The most significant contribution of technology has been the strengthening of real-time tailored engagement in various events like sports, music, etc.

2.4 *Creating Immersive Experiences*

Leading media platforms address and foresee consumer's personal entertainment requirements and drive an intense engagement by means of Artificial Intelligence—powered algorithms, playlists and recommendations. For instance in the popular science fiction anthology series "The Black Mirror" on Netflix, enables viewers to create their own narrative interpretations to the storyline resulting in a number of possible conclusions. Brodie et al. (2013) in their study relate consumer engagement to their interactive experiences. The engagement practices are supported by Tools like Virtual Reality (VR) and Augmented Reality (AR) for adaptive storytelling and immersive viewer experiences. These leverage advanced computer-aided techniques to dynamically redesign content into immersive digital formats and deliver unique experiences.

Virtual Reality is evolving as a tool for innovation and engagement in digital entertainment especially in the online gaming domain. Using a VR headset like Oculus Rift VR is a pragmatic and immersive simulation of a three-dimensional setting which enables an engaging experience in a computer-generated virtual environment. The tool is gaining momentum for its ability to radically improve customer experience. Virtual Reality is gradually marking its presence in the OTT space and creating innumerable avenues of consumer engagement and content distribution (Ahir, 2019). Augmented reality (AR) is yet another aggressively emerging tool enabling immersive experiences and engaging consumers in innovative ways that allows real-time view of elements, persons, or spaces in the material domain. The AR content can be made available virtually in different formats like images, videos, text and animation (Scholz & Smith, 2016).

Some of the most popular online games like X Box, Pokémon G generously leverage Augmented Reality (AR) by adding layers and digital elements for creating a fantastical experience to engage consumers. Snapchat among social media apps has been the pioneer in launching the concept of stories and spearheading the AR transition in entertainment. (Talebot, 2019) thus generating a sense of togetherness, camaraderie enhancing interactivity and participation. Findings of a survey conducted by Afaq news Bureau (April, 2019) indicate that with the rising millennial population and the subsequent digitalization especially in India, online social media networking sites, review sites and video platforms are evolving as crucial drivers for consumer interactions, This has resulted in an increased engagement with content on multiple wearable devices, and thus an experimentation with immersive content formats for generating relatable experiences.

2.5 The Role of Engaging Content

a. **Facilitation**

Content plays an important role in consumer's search for information, facilitating participation and thus creating consumer involvement. The ever growing telecom sector and the subsequent growth of cell phones has established a strong footing as the primary channel for content distribution and consumption with entertainment apps driving major consumer engagement. Professional believe that the engagement and interaction of content with young consumers is essentially motivated by their distinctive behavior and curiosity to explore, to be mobile and also socially visible by sharing (Ahir, 2019). The growing curiosity and need of young viewers for engaging content has further fueled concepts like Fear of missing out or FoMo (Auverset, Billings, & Conlin, 2016) and therefore evolution of novel delivery approaches like digital streaming giving them an easy access to all their most preferred content in a large quantity, without the anxiety or guilt of missing out. Literature also focuses on binge-watching (Spak, 2013), or accessing various episodes of content in continuation, through platforms like digital streaming or the act of wrapping up numerous episodes of a show in one go ("Netflix declares," 2013, para. 2).

The easy access to various channels along with the ability to consume and choose content anytime and anywhere within no time has created a sense of empowerment for viewers (Srinivasan, 2018). The plentiful alternatives and seamless ease of access facilitating content consumption on the go, like binge-watching further fueled the crucial role of content in consumer engagement. With a highly empowered and digitally democratic consumer personalized content can be created on various platforms by targeting the right mix of viewers based on their profiles and viewing habits, enabling robust customer-driven connections.

Studies conducted by different researchers acknowledges the wide spectrum and broad gamut of content delivered through various channels digitally ranging from newsletters, podcasts, live streaming videos, interactive quizzes, downloadable content info graphics, virtual conferences, webinars and a huge amount of freely available content for consumers who subscribe (Harris 2017; Taylor, 2012; Viswanathan et al., 2017).

b. **Over The Top (OTT) Players**

The digital consumers of today are grabbing media prospects to engage in content which is exclusively customized to their own personal experiences, backgrounds and agendas. Consumers with traits like high mobility and an extremely active disposition globally are highly empowered and therefore have a better control on how and when to utilize their media experience. Consumers are also enabled to manage media consumption, curate their own content assortment and get more digital content to their lives via a gamut of devices like smart phones, connected cars and platforms like OTT services.

Studies connect the trend of online video streaming and emergence of players like YouTube and Hotstar to this ease of accessibility to content of preference. (Srinivasan, 2018). Several new trends in internet browsing and a gradual change in social media browsing habits of the consumers could be observed due to the evolution of digital content. The demand for original content resulted in the rise to over the top (OTT) content consumption and engagement through mobile apps on smart and mobile devices. An outlook—PWC global report estimates OTT videos revenues to jump from 18% of global TV subscription revenue in 2018 to substantial 35.4% revenues in India alone.As per Boston Consulting Group analysis lucrative investments are being made on OTT content across platforms in various types of content with different offers and propositions (Samtani and Jindal, 2018) (Fig. 1).

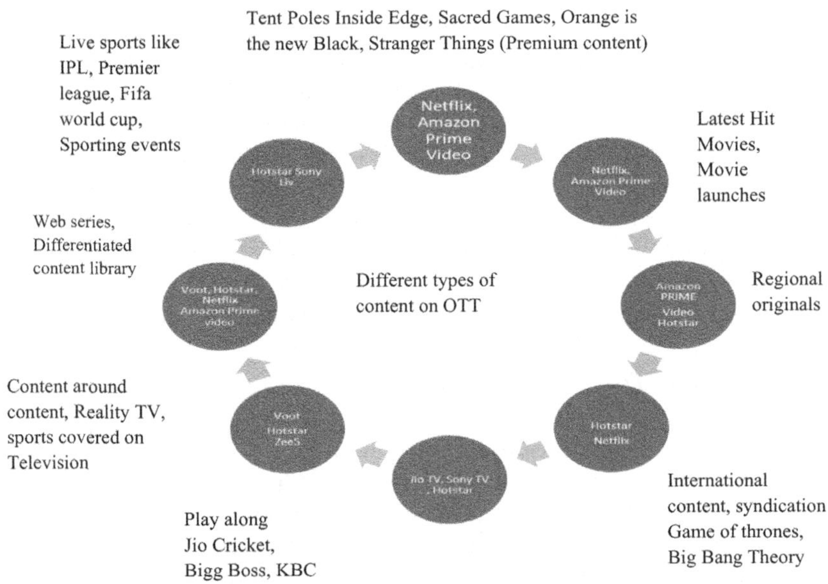

Fig. 1 Different platforms of digital entertainment (*Source* Adopted from Boston Consulting Group Report, November, 2018)

c. **User-Generated Content**

The current digital consumers are not only keen to experiment with different type of content but they also enjoy different genre of content. One way in which they engage digitally is by participating in co-creating content. Daugherty, Eastin, and Bright (2008) describe user generated content as the one which is not created by involving paid practitioners but is produced and delivered online by common people. While experience plays a critical role, digital content is also impacted by the consumer's contribution learning and co-creation in the experience as well as the partners and collaborators in the value chain (Rowley, 2008).

Facebook for instance allows consumers to participate by posting texts, images and videos, and also allows them to interact with others through features such as chats, comments, registering likes and even giving personal messaging through FB messenger. Twitter on the other hand has features that allows fans among consumers to connect with celebrities. By reinforcing the connection among fans, celebrities and content, the platform enhances user engagement (Samatani & Jindal, 2018).

Literature reveals that consumers not only enjoy reading content on social platforms but also seek gratification on getting comments for their posts further building interactions and taking the conversation forward by responding to the posts of others (Vargo & Lusch, 2004). (Dessart et al., 2016). One more engaging trait observed in today's digital viewers is their inclination to share relatable content with which they can connect (Maslowska et al., 2016). Dessart et al. (2016) review that the sharing-related trait is commonly observed on social platforms and also online brand communities as they are actually propelled by content.

While sharing on social platforms can be indicated by the comments, posts re-sharing via platforms like Face Book or as tweets re tweets and direct responses on Twitter (Gensler et al., 2013). Writing reviews, recommendations, writing blogs and spreading a positive word of mouth are a part of engaging with brand-related user generated content. (Hollebeek, Srivastava, & Chen, 2019; Yi, Gong, & Lee 2013).

3 Model Adaptation

With the above observations in literature many of the existing assumptions related to consumer engagement seem outdated creating a gap. This adds to the relevance and scope for the current chapter. Following are the major aspects that can be derived from the above review:

1. The enabling role of Technology in generating immersive experiences for consumers
2. Increasing role of social media in building relationships
3. Interactivity and Versatile application of content across platforms
4. Enhanced Consumer Involvement control and participation in creating content.

In the following section the author makes an attempt to adapt the Consumer Engagement Engine (Collinger et al., 2011) and apply it to the above discussion. The framework addresses in Fig. 2 and Table 1, the need for empowering consumers to create content and share their experiences. It recommends observation, participation and co-creation as significant indicators for an enhanced engagement and relevance. Adapting to the said Engine will fuel engagement further by inducing increased action.

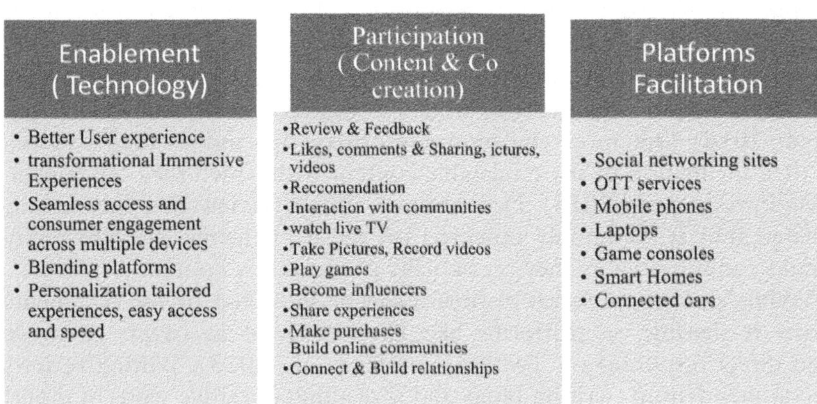

Fig. 2 Parameters of adoptive model (*Source* Compiled by the author)

Table 1 Summary of engaging digital entertainment

Digital Engagement tools	*Practical Applications*	*Consumer Engagement options*
Social media and networking sites (Facebook, Twitter, Instagram) **Interaction Type:** Personal and customized Networking and collaborative	Internet-based applications constructed on the fundamentals of Web 2.0 facilitating exchange and creation of user generated content (Kaplan and Haenlein (2010) Accessibility on Multiple devices Omnichannel: Apps, Sites Global appeal Versatile application Platform-based content Very High involvement participation and engagement e.g.,: Videos and pics **(Instagram)** (BARC, 2018) Huge personalized long conversations and people connect **(Facebook)** Crisp Ideas and notions with word limit and a call to action (feedback)—Engagement with celebrities **(Twitter)**(Breidbach and Brodie, 2014)	**Building Personal connections and networking through** Discussions Chats and interactions Friend requests Information events and nostalgia sharing and exchange Recommendations **Facilitating User generated content** Engaging Posts, blogs, reviews, vlogs, opinions, Feedback Entertaining videos, images, pictures, audio clips Individually owned pages **Unique Entertaining Content** Peek into Celebrity lifestyle and happenings Engagement with influencers, opinion leaders and celebrity. **Engaging Branded Content (Freemium & Premium)** TVCs Interactive Brand links Brand launches Interactive AR\VR-based content Brand display and purchase options Chat Bots

(continued)

Table 1 (continued)

Digital Engagement tools	*Practical Applications*	*Consumer Engagement options*
• Interactive digital entertainment (Online gaming, gambling and contests) • Zynga.com (Poker, Baceret), Zapak.com, Electronic arts.com, PUBG.com • Games: Rummy circle, **Interaction Type:** Recreational, knowledge based, engaging, addictive and entertaining	Engaging and enjoyable recreational activities with a transformational effect on leisure time Boyle, Connolly, Hainey, and Boyle (2011) Compatibility on multiple devices and platforms Like PCs, laptops, tabs, Smartphone's, consoles Intriguing use of augmented and virtual reality for a transformational effect Universal appeal across age groups, multiple user engagement avenues along with monetisation	**Skill-Based Games (Free & Premium)** Engaging Fantasy sports, Horse racing, **Non Skill-Based Games of Chance: (Free & Premium)** Casinos, Regulated gambling, lotteries, Card games like Rummy cards.
Videos on Demand, Mobile-based apps and streaming services Netlix, Hotstar, itunes, Tatasky, Amazon Prime, YouTube, YouTube premium	**Videos on demand service** (VOD) subscription-based interactive **technology** where consumers can access unlimited **videos on demand** and watch it on real time or at their preferred time by downloading the content Can be delivered across different platforms the internet to home personal computers, Satelite—Direct to home apps on Smartphones Over the top platforms and advanced digital devices	Versatile options available from music to sports Streaming delivery service (Live online Video streaming real time) Anytime anywhere accessibility through different delivery modes like mobile apps Overcome fear of missing out by binge-watching **Enjoy videos through Live streaming in different genres** Music Cricket Movies Original series

Digital Engagement tools	*Practical Applications*	*Consumer Engagement options*
		User generated content Create own videos of choice e.g.,: Recipes, Musical performances, Beauty and health tips, etc. Educational and motivational Share it through various platforms **Choice of engagement platforms** **Transactional Videos on Demand** (TVOD) Access content on a pay- per-view basis similar to an ala carte menu of content **Advertising Videos on Demand** (AVOD) Content can be accessed free with sponsored ads in between that cannot be skipped. **Hybrid Videos on Demand (HVOD)** Offer the consumers the choice to take up a combination of AVOD and SVOD. Or TVOD and SVOD (Flate rate plus some premium for specific content)

Source Compiled by the author

4 Conclusion

Digital viewers like to experiment with varied forms and genre of content. This chapter highlights the enabling role of technology in bringing about a seamless and engaging user experience across varied channels. It also tries to delve into the crucial role of content in facilitating consumer participation and co-creation. User generated content and engagement with brands is a growing trend in the digital entertainment space. The exponential rise of OTT players has opened newer avenues of engagement and entertainment. Consumers are always on the look-out for relatable content. Impactful content and innovative technology will propel the future of engagement and entertainment to the next level. However, despite the new norms in entertainment with ever evolving technology and consumer engagement, the concern is the growing addiction to digitals platforms and hence a near virtual socialization that might subsequently lead to isolation and mental depression. That apart with information and content becoming personalized the challenge going forward will be regulation of security privacy and protection of consumer data.

References

Abdul-Ghani, E., Hyde, K., & Marshall, R. (2011). Emic and etic interpretations of engagement with a consumer-to-consumer online auction site. *Journal of Business Research, 64*(10), 1060–1066.

AFaqs News Bureau. (2019). *Consumers in India spend over 10 hours daily on devices, engaging with content: Adobe's Brand content survey*, April. https://www.afaqs.com/news/marketing/54938_consumers-in-india-spend-over-10-hours-daily-on-devices-engaging-with-content-adobes-brand-content-survey.

AFaqs News Bureau (2019). *India's entertainment and media industry to clock over Rs 451,373 crore by 2023: PwC Report*. June 6.

Ahir, K. (2019), H*ow-can //OTT trump the-next way of entertainment with virtual-reality*. https://medium.com/datadriveninvestor/how-can-ott-trump-the-next-way-of-entertainment-with-virtual-reality. Accessed date: 21 March 2020.

Alvarez, C., & Fournier, S. (2016). Consumers' relationships with brands. *Current Opinion in Psychology, 10* 129–135.

Appel, G., Grewal, L., Hadi, R., & Stephen, A. T. (2020). The future of social media in marketing. *Journal of the Academy of Marketing Science, 48*(1), 79–95.

Armano, D. (2008). Brand interactions are the future: Be prepared with an interaction designer.*Advertising Age, 79*(17), 24.

Ashley, C., Noble, S. M., Donthu, N., & Lemon, K. N. (2011). Why customers won't relate: Obstacles to relationship marketing engagement. *Journal of Business Research, 64*(7), 749–756.

Auverset, L., Billings, A., & Conlin, L. (2016). Time-shifting vs. appointment viewing: The role of fear of missing out within TV consumption behaviors. *Communication & Society, 29*(4), 151–164.

Avery, E., Lariscy, R., Amador, E., Ickowitz, T., Primm, C., & Taylor, A. (2010). Diffusion of social media among public relations practitioners in health departments across various community population sizes. *Journal of Public Relations Research, 22*(3), 336–358. https://doi.org/10.1080/10627261003614427.

BARC. (2018). *What young India watches*? Retrieved from http://www.barcindia.co.in/resources/What%20Young%20India%20Watches.

Boyle, E., Connolly, T. M., & Hainey, T. (2011). The role of psychology in understanding the impact of computer games. *Entertainment Computing, 2*(2), 69–74.

Braojos-Gomez, J., Benitez-Amado, J., & Llorens-Montes, F. (2015). How do small firms learn to develop a social media competence? *International Journal of Information Management, 35*(4), 443–458.

Breidbach, C., & Brodie, R. (2014). Beyond virtuality: From engagement platforms to engagement ecosystems. *Managing Service Quality, 24*(6), 592–611.

Brodie, R., Ilic, A., Juric, B., & Hollebeek, L. (2013). Consumer engagement in a virtual brand community: An exploratory analysis. *Journal of Business Research, 66*(1), 105–114.

Brodie, R., Hollebeek, L., Juric, B., & Ilic, A. (2011). Customer engagement: Conceptual domain, fundamental propositions & implications for research in service marketing. *Journal of Service Research, 14*(3), 252–271.

Challagalla, G., Venkatesh, R., & Kohli, A. (2009). Proactive postsales service: When and why does it pay off? *Journal of Marketing, 73*(2), 70–87.

Cheung, C., Lee, M., & Jin, X. (2011). Customer engagement in an online social platform: A conceptual model and scale development. In *ICIS 2011 Proceedings* (Paper 8). Shanghai, China.

Collinger, T., Malthouse, E. C., & Maslowska, E. (2011). *The consumer engagement engine*. Retrieved from spiegel.medill.northwestern.edu.

Das, S. (2020). Innovations in digital banking service brand equity and millennial consumerism. *Digital transformation and innovative services for business and learning* (pp. 62–79). IGI Global. https://doi.org/10.4018/978-1-7998-5175-2.ch004.

Das, S. (2021). *Search engine optimization and marketing: A recipe for success in digital marketing* (1st ed.). CRC Press. https://doi.org/10.1201/9780429298509.

Das, S., & Mondal, S. (2016). *Yours digitally: Patanjali*. Available at SSRN 2737299.

Das, S., & Nayyar, A. (2019). Innovative ideas to manage urban traffic congestion in cognitive cities. In *Driving the development, management, and sustainability of cognitive cities* (pp. 139–162). IGI Global. with https://doi.org/10.4018/978-1-5225-8085-0.ch006.

Das, S., & Nayyar, A. (2020). Effect of consumer green behavior perspective on green unwavering across various retail configurations. In *Green marketing as a positive driver toward business sustainability* (pp. 96–124). IGI Global. with https://doi.org/10.4018/978-1-5225-9558-8.ch005.

Das, S., Nayyar, A., & Singh, I. (2019). An assessment of forerunners for customer loyalty in the selected financial sector by SEM approach toward their effect on business. *Data Technologies and Applications, 53*(4), 546–561. https://doi.org/10.1108/DTA-04-2019-0059.

Das, S., Mondal, S. R., Sahoo, K. K, Nayyar, A., & Musunuru, K. (2018). Study on impact of socioeconomic make up of Facebook users on purchasing behavior. *Revista Espacios, 39*, 28–42. Retrieved from http://www.revistaespacios.com/a18v39n33/18393328.html.

Daugherty, T., Eastin, M. S., & Bright, L. (2008). Exploring consumer motivations for creating user-generated content. *Journal of Interactive Advertising, 8*(2). Retrieved from http://jiad.org/article101.html.

Davis, F. D. (1989). Perceived usefulness, perceived ease of use, and user acceptance of information technology. *MIS Quarterly, 13*, 319–339.

Dessart, L., Veloutsou, C., & Morgan-Thomas, A. (2016). Capturing consumer engagement: Duality, dimensionality and measurement. *Journal of Marketing Management, 32*(5–6), 399–426.

Duy, N. T., Mondal, S. R., Van, N. T. T., Dzung, P. T., Minh, D. X. H., & Das, S. (2020). A study on the role of web 4.0 and 5.0 in the sustainable tourism ecosystem of Ho Chi Minh City, Vietnam. *Sustainability, 12*(17), 7140. https://doi.org/10.3390/su12177140.

Garretson, R. (2008). *Future tense: The global CMO*. Available at http://graphics.eiu.com/upload/Google%20Text.pdf. Accessed on 14 April 2020.

Gensler, S., Völckner, F., Liu-Thompkins, Y., & Wiertz, C. (2013). Managing brands in the social media environment. *Journal of Interactive Marketing, 27*(4), 242–256.

Gupta, D. K., Jena, D., Samantaray, A. K., & Das, S. (2019). HRD climate in selected public sector banks in India. *Revista ESPACIOS, 40*(11), 14.

Hardaker, S., & Fill, C. (2005). Corporate service brands: The intellectual and emotional engagement of employees. *Corporate Reputation Review, 7*(4), 365–376.

Harris, J. (2017). *Cover all the bases with 21 winning content marketing techniques*. at: http://contentmarketinginstitute.com/2017/05/winningcontent-marketing-tactics/. Accessed 21 March 2020.

Hollebeek, L. D., Srivastava, R. K., & Chen, T. (2019). SD logic–Informed customer engagement: Integrative framework, revised fundamental propositions, and application to CRM. *Journal of the Academy of Marketing Science, 47*(1), 161–185.

Hollebeek, L. D., Glynn, M. S., & Brodie, R. J. (2014). Consumer brand engagement in social media: Conceptualization, scale development and validation. *Journal of Interactive Marketing, 28*(2), 149–165. https://www.ey.com/en_in/media-entertainment/ficci-ey-media-and-entertainment-report-2019.

Hughes, P. (2019). *Digital rules of (consumer) engagement*. Retrieved from https://www.multichannel.com/blog/digital-rules-of-consumer-engagement.

Jain, S., Jain, V., & Das, S. (2018). Relationship analysis between emotional intelligence and service quality with special evidences from Indian banking sector. *Revista ESPACIOS, 39*(33), 3.

Järvinen, A., Heliö, S., & Mäyrä, F. (2002). *Communication and community in digital entertainment services*. Prestudy Research Report.

Kaplan, A. M., & Haenlein, M. (2010). Users of the world, unite! The challenges and opportunities of Social Media. *Business horizons, 53*(1), 59–68.

Katiyal, T. (2020). *In an OTT world, only the consumer is on top*. Retrieved from https://brandequity.economictimes.indiatimes.com/be-blogs/media/3977.

KPMG. (2019) *Unravelling the digital video consumer: Looking through the viewer lens*. Retrieved from https://home.kpmg/in/en/home/insights/2019/09/ott-digital-video-market-consumer-india.html.

Kumar, V. K., & Gupta, S. (2016). Conceptualizing the evolution and future of advertising. *Journal of Advertising, 45*(3), 302–317.

Kuwamura, Y. (2020). *Digital trends set to transform how Indians consume content*. https://brandequity.economictimes.indiatimes.com/news/digital/digital-trends-set-to-transform-how-indians-consume-content-in-2020/74526604.

Lamberton, C., & Stephen, A. T. (2016). A thematic exploration of digital, social media, and mobile marketing: Research evolution from 2000 to 2015 and an agenda for future inquiry. *Journal of Marketing, 80*(6), 146–172.

Maslach, C., & Leiter, M. P. (1997). *The truth about burnout: How organizations cause personal stress and what to do about it*. San Francisco, CA: Jossey-Bass.

Maslowska, E., Malthouse, E. C., & Collinger, T. (2016). The customer engagement ecosystem. *Journal of Marketing Management, 32*(5–6), 469–501.

Mohanty, P. C., Dash, M., Dash, M., & Das, S. (2019). A study on factors influencing training effectiveness. *Revista Espacios,* 40, 7–15. Retrieved from http://www.revistaespacios.com/a19v40n02/19400207.html.

Mondal, S. R. (2020). A systematic study for digital innovation in management education: An integrated approach towards problem-based learning in Vietnam. In *Digital innovations for customer engagement, management, and organizational improvement* (pp. 104–120). IGI Global.https://doi.org/10.4018/978-1-7998-5171-4.ch006.

Mondal, S., & Sahoo, K. K. (2020). A study of green building prospects on sustainable management decision making. In *Green building management and smart automation* (pp. 220–234). IGI Global. https://doi.org/10.4018/978-1-5225-9754-4.ch011.

Mondal, S., Das, S., Musunuru, K., & Dash, M. (2017). Study on the factors affecting customer purchase activity in retail stores by confirmatory factor analysis. *Revista Espacios, 38*(61), 30.

Mondal, S., Mall, M., Mishra, U. S., & Sahoo, K. (2017). Investigating the factors affecting customer purchase activity in retail stores. *Revista Espacios, 38*(57), 22.

Nadanyiova, M., & Das, S. (2020). Millennials as a target segment of socially responsible communication within the business strategy. *Littera Scripta, 13*(1), pp. 119–134. https://doi.org/10.36708/Littera_Scripta2020/1/8.

Netflix. (2013, December 13). *Netflix declares binge-watching is the new normal. [Press release].* Retrieved September 2015, from https://pr.netflix.com/WebClient/getNewsSummary.do?newsId=496.

Perspectives from Global Media and Entertainment Outlook 2019–2023. *Getting personal: Putting the Me in entertainment and media.* Retrieved from https://www.pwc.com/gx/en/entertainment-media/outlook-2019/entertainment-and-media-outlook-perspectives-2019-2023.pdf.

Pherwani, A. (2020). *How will the M&E sector navigate the era of consumer A.R.T.?* Cambridge, MA: Academic Press.

Rowley, J. (2008). Understanding digital content marketing. *Journal of Marketing Management, 24*(5–6), 517–540.

Roy, D. (2019). *Importance of consumer-engagement for brand retention in the digital-age.* https://brandequity.economictimes.indiatimes.com/news/marketing/importance-of-consumer-engagement-for-brand-retention-in-the-digital-age/72215852. Accessed 21 March 2020.

Samtani, K., & Jindal, G. (2018). *Entertainment goes online A $5 billion opportunity.* https://images.assettype.com/afaqs/import/all/news/images/news_story_grfx/2018/11/53972/BCG-Entertainment%20Goes%20Online".pdf.

Schaufeli, W., Salanova, M., Gonza´lez-Roma´, V., & Bakker, A. B. (2002). The measurement of engagement and burnout: A two sample confirmatory factor analytic approach. *Journal of Happiness Studies, 3,* 71–92.

Scholz, J., & Smith, A. N. (2016). Augmented reality: Designing immersive experiences that maximize consumer engagement. *Business Horizons, 59*(2), 149–161. https://doi.org/10.1016/j.bushor.2015.10.003.

Sharma, E., & Das, S. (2020) Measuring impact of Indian ports on environment and effectiveness of remedial measures towards environmental pollution. *International Journal of Environment and Waste Management, 25*(3), 356–380. https://doi.org/10.1504/ijewm.2019.10021787.

Singh, I., Nayyar, A., & Das, S. (2019). A study of antecedents of customer loyalty in banking & insurance sector and their impact on business performance. *Revista ESPACIOS, 40*(06), 11.

Singh, I., Nayyar, A., Le, D. H., & Das, S. (2019). A conceptual analysis of internet banking users' perceptions. An Indian perceptive. *Revista ESPACIOS, 40*(14), 1.

Singh, L. B., Mondal, S. R., & Das, S. (2020). Human resource practices & their observed significance for Indian SMEs. *Revista ESPACIOS, 41*(07). Retrieved from http://www.revistaespacios.com/a20v41n07/20410715.html.

Singh, S., & Das, S. (2018). Impact of post-merger and acquisition activities on the financial performance of banks: A study of Indian private sector and public sector banks. *Revista Espacios Magazine, 39*(26), 25.

Singh, S., Mondal, S., Singh, L. B., Sahoo, K. K., & Das, S. (2020b). An empirical evidence study of consumer perception and socioeconomic profiles for digital stores in Vietnam. *Sustainability, 12*(5), 1716.

Spak, K. (2013, December 13). *Half of Netflix bingers watch a season in a week* [Web log post]. Retrieved from http://www.newser.com/story/179135/half-of-netflix-bingers-watch-aseason-in-a-week.html.

Srinivasan, A. (2018). *DTH vs streaming services: Is DTH nearing its demise in India?* The Mobile Indian. Retrieved from https://www.themobileindian.com/news/dth-vs-streaming-services-is-dth-nearing-its-demise-in-india-23414.

Talbot, K. (2019). *How-snapchat-is-transforming-the-physical-world-through-augmented-reality.* https://www.forbes.com/sites/katetalbot/2019/05/01/how-snapchat-is-transforming-the-physical-world-through-augmented-reality/#a0681a44e4c1. Accessed 22 March 2020.

Taylor, G. (2012). *Digital Content Marketing*. Scotts Valley, CA, USA: CreateSpace Independent Publishing Platform.

Van, N. T. T., Vrana, V., Duy, N. T., Minh, D. X. H., Dzung, P. T., Mondal, S. R., & Das, S. (2020). The role of human–machine interactive devices for post-COVID-19 innovative sustainable tourism in Ho Chi Minh City, Vietnam. *Sustainability, 12*(22), 9523. https://doi.org/10.3390/su12229523.

Van Doorn, J., Lemon, K. N., Mittal, V., Nass, S., Pick, D., Pirner, P., & Verhoef, P. C. (2010). Customer engagement behavior: Theoretical foundations and research directions. *Journal of Service Research, 13*(3), 253–266.

Vargo, S. L., & Lusch, R. F. (2004). Evolving to a new dominant logic for marketing. *Journal of Marketing, 68*(1), 1–17.

Vogt, C., & Knapman, S. (2008). *The anatomy of social networks.*

Viswanathan, V., Hollebeek, L., Malthouse, E., Maslowska, E., Kim, S., & Xie, W. (2017). The dynamics of consumer engagement with mobile technologies. *Service Science, 9*(1), 36–49.

Vivek, S. D., Beatty, S. E., & Morgan, R. M. (2012). Consumer engagement: Exploring consumer relationships beyond purchase. *Journal of Marketing Theory and Practice, 20*(2), 122–146.

Yi, Y., Gong, T., & Lee, H. (2013). The impact of other customers on customer citizenship behavior. *Psychology and Marketing, 30*(4), 341–356.

CHAPTER 3

A Study on Processing of Information Storage & Use of New Age Consumers in Digital Wellness Sector Through Story Telling & Creating Interest

Ramachandran Siri and Subhankar Das

1 Introduction

The consumer information processing model focusing on consumer behavior stages and explains the consumer's cognitive and operational consumption behaviors. The role of advertising in consumer information processing phase is also significant (Nadanyiova & Das, 2020; Singh, Mondal, Singh, Sahoo, & Das, 2020). Advertisements perform an important role in the information processing phase as part of health communication. For example, in some private television channels, advertising spots related to herbal creams for joint diseases are broadcasted

R. Siri
University of New England, Armidale, NSW, Australia

S. Das (✉)
The Honors Programme, Duy Tan University, Da Nang, Vietnam
e-mail: subhankardas@duytan.edu.vn

S. Das and S. Gochhait (eds.), *Digital Entertainment*,
https://doi.org/10.1007/978-981-15-9724-4_3

(Singh, Mondal, & Das, 2020). These ads, which are on screens for long minutes, are generally aimed the elders who complain of joint pain, and related creams are alleged to be effective in stopping many types of pain without distinction of its kind (Sharma & Das, 2020). These advertisements are coded according to the consumer information processing model (Das, 2020).

Related ads use a metaphorical language; the cream also takes advantage of a religious motive by claiming that it facilitates the prayers for those who have difficulty in bending to perform the ritual by removing the joint pains (Mondal & Sahoo, 2020). Therefore, in this study, herbal cream advertisements for joint diseases will be examined through advertisements of 3,000-year-old Miracle Cream, Oriental Medicine's Miracle Cream and Miracle Cream *(3.000 Yıllık Mucize Krem, Doğu Tıbbının Mucize Kremi* & *Mucize Krem*) which are broadcasted in the middle of 2017 (Das & Nayyar, 2020). The research will examine how the relevant advertisements are coded, connected and supported by the recipients complaining of joint disease. In this regard, the role of advertising in the context of consumer information processing model will also be discussed (Das, Nayyar, & Singh, 2019). Because the advertisements are a guide for the consumer to see the problem, to search information/solution about it, to see the alternative and to take action (Das & Nayyar, 2019). For this reason, in this study, the content and role of advertisements in the context of consumer information processing phase will be discussed through herbal cream advertisements selected as a sample and directed to joint pains (Singh, Nayyar, Le, & Das, 2019).

1.1 Consumer Information Processing Model and Advertisements

The present meaning of health communication notion is shaped by the perception of "supporting health" (Singh & Das, 2018). If this is not the case, health communication will only describe that the doctor gives the patient a good explanation about the treatment and useful pieces of information about the disease, and that the health practitioners also transfer the information to the public with a correct language and right channels (Mohanty, Dash, Dash, & Das, 2019). Yet, considering that communication is an interactive process, it will carry health communication notion to a wider frame (Erbaydar, 2003: 51). Health communication is very important for public health önemlidir (Erbaydar, 2003). At this point, it should be stated that a large number of studies in the field of health communication have found that communication campaigns held in this area meet the problems of reaching the target audience (Ergül, 2013: 166).

The consumer information processing model, which provides useful applications for the healthcare field, has not been developed to be applied in the field of health communication or to be examined health-related behaviors (Singh, Nayyar, & Das, 2019). However, it is a model as shown in Fig. 1, used in studies on health communication, and it is quite explanatory to understand the consumer behavior (Gupta, Jena, Samantaray, & Das, 2019). According to the consumer information processing model, the consumer, namely, the person, is required to identify problems, search for information, evaluate and select alternatives, apply the decision and evaluate it after the purchase (Okay, 2014: 87–89 cited from Thomas, 2006: 67).

The consumer, firstly, starts the search for internal or external information by finding out what the problem is (Jain, Jain, & Das, 2018). With information after the search, he identifies and evaluates the alternatives for the problem. Then the purchase process begins, even though this process can be thought as the end of consumer behavior, there is one more step. This step is the evaluation after the purchase. By making an evaluation after purchasing, the consumer interrogates and evaluates his actual

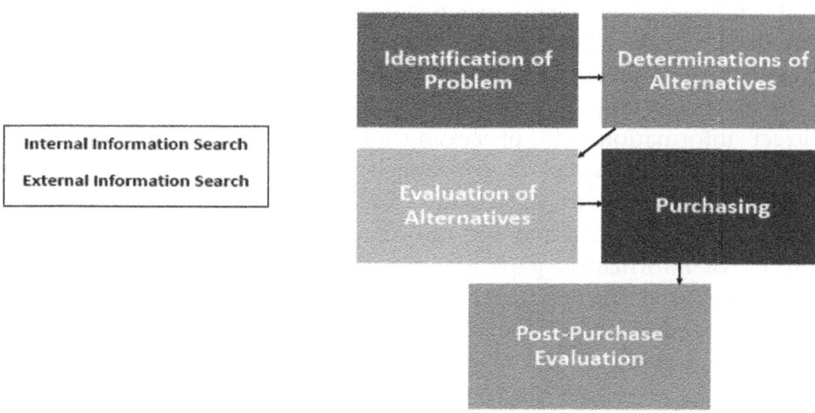

Fig. 1 The Place of the Information Search in the Purchasing Process (Cited in Altunışık & Baş, 2015: 83)

consumer behavior along with the product/service/goods (Altunışık and Baş, 2015; Yıldırım, 2016).

The high quality of information (information's nature, consistency or understandability), on the one hand, increased the decision performance by enabling the consumer to process this information faster and easier, on the other hand a lot of information that was with good quality or valuable to the consumer made it difficult for the consumer to take effective decisions (Keller and Staelin, 1987; Özata, 2011: 14). The consumer should be treated as a whole with both the cognitive processes in his inner world and the ways of behavioral solutions in his outer world. For this reason, the inner world of the consumer as an individual becomes the center of interest for the marketing researchers (Altunışık and Baş, 2015: 80). The relationship between consumers' demographic characteristics and purchasing styles is also important. It is known that "Consumer Buying Styles" (Consumer Type Inventory CTI), which is based on the assumption that consumers make purchasing decisions according to certain styles, is assumed to be a powerful measurement tool (Erdem, 2016). Factors affecting consumer purchasing behavior are generally cultural, personal, social and psychological (Cömert and Durmaz, 2006). "Brand loyalty" is also important in the purchasing decision process (Fırat and Azmak, 2007). The study was conducted to determine brand loyalty through *Toyota* car users showed that brand image and loyalty was highly insignificant with age group, level of education, point of purchase and repeat purchase duration (Das, 2021; Duy et al., 2020; Shafiq Gul, Ali Jan, Baksh Baloch, Faheem Jan, & Farooq Jan, 2010; Van et al., 2020).

In the marketing strategies activity, the importance of accessing the correct information and processing of information is obvious (Das, Mondal, Sahoo, Nayyar, & Musunuru, 2018). Especially the increase of information due to the changes and developments experienced in the field of communication and information technologies caused a large number of information gaps between the business and the consumers and necessitated the reorganization in the marketing practices because of the increasing knowledge with the developing technology (Aktuğlu, 2006). As a result of the revolution in computer and communication technologies, important changes have taken place in economic life as well as in social life (Dilek, 2016: 90). Information has got ahead traditional production factors in recent years and has become the most important production factor in many sectors (Avcı and Avcı, 2004). The Internet is extremely important for marketing. Studies evaluating the consumer purchasing decision process in marketing activities on the Internet have also been carried out (Özcan, 2010).

New media environments today have brought a different dimension to health information and health communication (Mondal, Das et al., 2017). Television and internet are the most preferred two media for health-related issues. The researches on how to use media technologies that gain importance day by day in the field of health communication are becoming more and more important (Hülür, 2015: 28). The underlying reason for the studies carried out in the field of health communication is the principle of raising awareness by informing and directing individuals of the society toward correct behavior. In this context, parallelly to the growing power and importance of the new media, organizations operating in the healthcare field should follow social networks and use them efficiently (Mondal, Mall et al., 2017). Because they will also have knowledge about current social events and practices. In addition to this, it should not be forgotten that social media tools such as *Facebook* and *Twitter* stand out today as interactive platforms in terms of news/information obtaining as well as news/information sharing (Öztürk and Öymen, 2014: 127).

According to Kara (2017: 12), the most basic criteria in social media as part of health communication should be to disseminate valuable health information, to increase target groups, to personalize messages, to ensure participation and interaction. The underlying reason for the studies carried out in the field of health communication, is the principle of raising awareness by informing and directing individuals of the society toward correct behavior (Das & Mondal, 2016). In this context, developing communication technologies should not be left behind. Therefore, in a study, Kara (2017), who examined how and for what purposes hospitals use the popular social media application Instagram in the context of health communication, analyzed the sharings of hospitals. These kinds of studies reveal how and in what extent health communication is profited in new media environments.

In the consumer purchasing process, advertisements serve as an important guide. The consumer who has started the search for information after knowing the problem (at this point the information is the raw data as a whole, and the information is the processed version of the information) can access the information through advertisements, see and evaluate the alternatives.

2 Marketing vs. Advertising Review

A general definition is made for the advertisement: "Advertising makes the promotion of goods and services to large audiences through media such as magazine, radio, television, poster." (And now through media too…) Advertising is a tool that informs the consumer about where the product is sold, price and quality, use of the goods and services in question, and shows him the best way to make good use of his money. In addition, advertising is a great support factor for the business man to find a good market, evaluate his capital and effort, and focus on new production and investments (Pektaş, 1987: 221). The advertising, which aims to motivate the feelings that are claimed to be the result of ideas and invites the consumers to consume, is "the use of communication channels by a defined sponsor or organization paying the price to reach large masses." The target group of the advertising is the buyer in other words customer/consumer. The customer is "the person who buys and uses an organization's products or services" (Heath and Nairn, 2005; Karlsson, 2007: 7–8). Advertising is the most important and effective medium of marketing a product or service. There is a long history of using advertising for marketing purposes In fact, while the consumer's emphasis on consuming more than their needs exposes the consumer society, advertisers have not delayed their adoption of a strategy to use the inadequacies of consumers. Thus, products are being marketed to ease the negative self/missing self image of the consumer (Calvert, 2008: 205). Advertising is also a form of communication. However, one of the most important things that distinguish advertising from other forms of communication is the promotion of observation (Karimova, 2014).

As a marketing argument, advertising plays an important role in our daily lives today. Advertisements mainly identify the image and lifestyle. Ads also have an impact on our attitudes toward ourselves and our world around us. The advertising shows the forms on-hand for behavior in a specific way. It determines what is good, what is bad, and in fact, everyone is involuntarily impressed by advertising. Nowadays, as we become a slave to the advances in science and technology, advertisements use the progressions efficiently (Chan, 2017; Frolova, 2014). On the other hand, the way to make undoubtedly an effective and great advertisement; is to plan, to create a concept and effortfulness (Gözler, 1999). Creativity is also important in forming an effective advertising and provides an in-depth look at the process of composing ideas and concepts. After all, there are various ways to produce ideas and encourage creativity (Collins, 2014; Hopkins, 2015; Mahon, 2011).

The effectiveness of advertising messages is widely believed to be based on audience participation (Elden, 2016; Cialdini, 2017). There are four levels of participation in this context; preattention, focal (attention), understanding and elaboration (Arvidsson, 2006). These levels dedicate increased attention capacity to a message source, as needed for the analysis of the message by different representative systems, increasingly abstract and qualitatively (Greenwald & Leavitt, 1984: 581; Mackay, 2005). For this reason, in order to prepare a good advertisement, it is necessary to better understand consumer behavior and expectations (Greenwald & Leavitt, 1984). It is also necessary to avoid the use of exaggerated and difficult to realize expressions in the ad text (Güler İplıkçı, 2015). Such statements may cause doubts about the product or undermine the trust in the product and the company. Likewise, over-selling air in the text decreases the prestige of the product and the company and awakens the image of the product with low quality (Bozgeyik, 2013). It should also be stated what benefit the consumer will get by using or using the advertised product or service in the text (Maya TV, 2017). Because we know that people tend to behave according to their interests (Kasım, 2004: 143).

A good advertising agency does business with systematic planning, research and intense effort with its numerous units (Karabaş, 2013). In an advertising agency consisting of certain sections, the creative group examines the request of the customer, prepares a text and an appropriate draft is drawn up for the selected media (Pektaş, 1987: 224). It is claimed that one of the former American presidents, Franklin Roosevelt, said: "*If I had the opportunity to start life again, I would have preferred advertising without any hesitation.*" According to this claim, Winston Churchill said, "*The consumption power of people is the food of advertising. It creates a desire for the best living standards. It instills in people the best nutrition, the best dressing, the best homes for themselves and their families.*" So indeed, these approaches emphasizes the importance of advertising (Pektaş, 1987: 224).

According to Phillips (2006: 92), "other tools will be used if advertising is withdrawn from reaching capitalist goals. However, it is unlikely that we can attack the root of the problem by eliminating capitalism. No viable alternative appears in the modern world. In addition, such an idea is defenseless; because its correction can create more social problems than expected. We live in a good, bad, capitalist economic system and we have to work in it to solve problems" (Phillips, 2006: 92). According to Phillips (2006), "ads have been criticized ever since they appeared. Critics tend to get stuck in the best-known argument against advertising. According to this argument, advertising manipulates or forces consumers to buy unnecessary or unwanted products. Their supporters oppose the idea that advertising plays an important role in providing information and cannot force skeptical consumers to buy a product they do not want to buy" (Phillips, 2006).

Advertising is also a major source of media financing (Ellman and Germano, 2008: 1). Television, in particular, attaches great importance to advertisements in terms of revenue (Shachar & Anand, 1998: 363). The advertising market that has vast sums (Goldstein, Mcafee, & Suri, 2012) surrounds individuals for so long. Despite the presence of those who seek ad-free "good old days," there is also a large audience that is intensively influenced by advertisements. This mass lead by youth consumers and brand addiction also manifests itself in social media today. In fact, social media today has become an advertising and consumer medium and especially the products advertised by celebrities are almost the main content of social media (Kwak et al., 2010). At the same time, there are many examples of creative and effective advertising that draw attention to social media as well as studies on how to advertise effectively in social media today (Pope, 2003; Yılmaz & Taşkıran, 2015).

Promoting their products on social media increases the recognition of brands (Mestçi, 2017). "Firms that want to strengthen their image and increase their sales want to take advantage of the blessings of both experiential and viral marketing by acting with new kind opinion leaders to provide customer experience. Brands that communicate with new generation leaders on social media platforms can reach countless people in a short time (Michaels, 2014). Thus, companies reduce their marketing budgets." An effective and interactive process is also mentioned here (Alikılınç & Özkan, 2018: 43).

Social media has changed with the advent of Youtube and its contextual usability (Mondal, 2020). With the concept of blog, people started to get ideas and advice online. In addition, the people who made new generation opinion leadership who directed the masses with the videos they shot enabled the profession of "vloggers" to emerge. People who recorded their lives through a camera began to offer it to the masses. These people also advertise and promote in their videos, and this is a current, remarkable and substantial argument for the advertising industry

(Reeves, 1961). In this context, studies that deal with the channels on Youtube within the framework of the vlog element in the new generation understanding of advertising are also carried out (Özdel, 2018). Carrying out these studies is also important in revealing the digital transformation in advertising. Because very clearly, we see that today many areas of human life and living spaces are experiencing digital transformations. Like the concrete or abstract consumption conceptions in human life, the reflections of these understandings are also transforming (Topsümer & Elden, 2016).

Advertising has a vital role in today's competitive marketing world. Advertising that helps sell goods and services is an essential and necessary tool for business marketing. Advertisements are also important for competition between businesses and affect product price and quality. It also serves as an assistant to accelerate the distribution of advertising innovations, the industrialization process and the expansion of businesses. Advertising also helps to increase the productivity of companies and raises social standards (Terkan, 2014: 240).

A successful ad can create a high prestige for the firm and the product or the opposite. A successful ad can create a high prestige for the firm and the product or the opposite. For instance, a successful ad prepared with intense efforts may increase the price of the products/services, while the price of the products/services may decrease if the ad is poorly prepared (Karimova, 2014), the firm may lose prestige.

The field of advertising is developing at an unprecedented proportion today. The field of advertising is developing at an unprecedented proportion today. Influenced by media fragmentation and population change factors, the field addresses intensely younger, more diverse, more technologically savvy consumers and knows how to reach them (Advertising & Audiences, State of the Media, 2014). Today, advertising often creates its desired effect by using soft effective and cohesive models that focus on persuasion (Vakratsas & Ambler, 1999: 28–33). At this point, while trying to figure out how and for what advertisements are done, it is also necessary to think about what was emphasized, anticipated or postponed (Williamson, 2002: 17).

2.1 *Sales Through the Internet and Possible Problems*

In parallel with the development of communication technologies day by day, a large number of new and different products are introduced and presented to the market. While these products are generally intended to facilitate the daily practices of individuals, they address individuals from a variety of occupations, age groups and genders. The advertisements benefited from many facts by considering the social structure and living conditions promise to facilitate life. For example, TV commercials directed toward women aim often to provide an aid to them on house/kitchen works, while commercials for men focus on hair and its products, auto accessories and construction equipment (saws, etc.). However, advertisements vary according to seasonal needs. For example, in some channels, commercials for products such as inflatable pool in summer and coat and boots in winter.

It is alleged that products sold today, especially over the internet or through some television commercials, often arrive at consumers differently than expected, so the consumer can be aggrieved. For instance, 73-year-old Huseyin Eninanç, who lives in Sivas, bought a knife set through a TV commercial that he watched to use on the feast of sacrifice, he was shocked when the knife that he paid for 99 TL (Turkish Lira), was arrived. Instead of the wooden handle knifes that he saw at the commercial, Eninanç confronted simple knives with plastic handles sold for 2–3 TL (Turkish Lira) on the market, alleged that he could not even cut a lemon with that knife. According to him, Eninanç had recalled the phone number but he could not reach the number again and could not find an authorized person to talk. It is noted that consumers are taken in with the imitated products of many firms especially the handmade Sürmene knives of Emrullah Başgüldekendi's firm (http://www.hurriyet.com.tr, August 24, 2017). After this incident, a firm that claims to sell handmade Sürmene knives has carried Eninanç on their commercial, to warn the consumer against fake products.

In recent years, a similar situation has been encountered by some women who practice shopping from a TV commercial they watched. It has been claimed that aluminum pans have been sent from the TV channel in fact they ordered steel cookers. Fraud activities are also detected in some sales over the Internet. Consequently, there are fraud activities in the advertising market through TV as well as internet. As a result, consumers are being warned by consumer associations, also by the media through the internet and some publications against fraud.

2.2 *Herbal Cream Ads for Joint Diseases*

In mid-2017, in some satellite TV channels, commercial spots related to creams mostly the herbal ones for joint diseases are broadcasted. These ads, which are quite long, usually address drivers, housewives and elderly people; it is promised to relieve the pain with the effect of herbal cream, regardless of its cause or type. For example, some of these creams are claimed to cure neck, waist and muscle pains all at once.

In addition to herbal creams marketed in some TV channels, some gifts for consumers/buyers are also promised. For example, it has been declared that besides a black sesame oily cream, (the price of 4 boxes is 49 TL [Turkish Lira] including taxes, shipping free) here the attention is drawn to its richness in vitamins B1, B2 and B6, the KIBLEGAH Islamic game set worth 49 TL (Turkish Lira) is to be presented as a gift. It has been alleged that the concerned cream helps to renew the skin texture and strengthen the underneath glands by protecting the skin from external influences.

Commercials, usually shown among comedy shows (Especially, Kemal Sunal, İlyas Salman and Şener Şen movies. The focus mostly on Kemal Sunal movies), is attempted a destination market through religion. A set of Islamic game as a gift besides the cream as shown in the above example, it is also promised to give a thermal most sock with the product in another black sesame cream commercial. Religion motif stands out as an additional marketing argument in ads that says "First Allah, then we will find a remedy to ourselves, by Allah's allow we will conduce to this."

As of October 2017, it is observed that some televisions advertised vegetable oils (such as Scythian oil) instead of herbal creams for joint diseases. The celebrities (Latif Doğan, Songül Karlı, İzzet Altınmeşe) that involved in these commercials are even the same ones. While the ad uses the slogan "a must for every household," it is also mentioned that "***this product is not a medicine.***" According to my observation, the product packaging and some scenes from the advertisements are similar to the previously marketed herbal creams and its contents.

"3,000-year-old Miracle Cream," "Miracle Cream of Oriental Medicine"* and *"Miracle Cream"

In the mid-2017, there were remarkable herbal cream advertisements aired on some television channels. One of these is *3,000-year-old Miracle Cream. 3,000-year-old Miracle Cream* which is advertised as buy 1, get 1 for free, is 59 TL (Turkish Lira) and its shipping is free.

In cream advertisement, mostly elderly people who are suffering from joint pain and famous people (Latif Doğan) are involved and they explain with the examples from their lives about their use and satisfaction with the cream. For example, an elderly woman says that before using the cream, she could not even climb up the stairs because of the pain in her knees but now she says she is relieved from the pain, while another person says she was bending for the pray due to her knee pain, but now she prays properly thanks to the cream. It indicated that in case of dissatisfaction the product can be returned. Besides, the person whose role is difficult to distinguish: the marketing agent or the TV presenter, makes a call to the consumers, "Call us if your neighbor bought the product and he/she is satisfied, call us if you see from your neighbor" The product claims to be effective in many pain and illnesses as back hernia, neck hernia, neck pain. Additionally, it is observed that the phone numbers in the ads are in different formats for different channels where the ad is shown.

Another herbal cream for joint diseases commercial *Miracle Cream of Oriental Medicine* is broadcasted at the same period with the *3,000-year-old Miracle Cream*. The price of the *Miracle Cream of Oriental Medicine* advertised as buy 1, get 1 for free, is 59 TL (Turkish Lira) with shipping included. On the front of the cream box, there are non-Turkish writings and a cat figure.

In *Oriental Medicine's Miracle Cream* advertising, elderly women and men and middle-aged men describe the effects of cream and the benefits they had from cream. According to their description, the cream reveals the pain which exists for years in a few minutes. That is why they recommend this product to everyone. There are also male and female hostesses dressed locally beside the TV presenter in an advertisement.

In the last example of the advertisement *Miracle Cream*, there is the same presenter and the price is 59 TL (Turkish Lira) with shipping included and again buy 1, get 1 for free as in the previous advertisement. From this point of view, the cream can be confused with *Oriental Medicine's Miracle Cream*. In fact, this ad and the cream differ in the name/brand. In the related ad, middle age is the leading role; the middle-aged women and men explain the effects and benefits of the cream, and support the presenter's claims about the cream effect. However, the number of people involved in this ad is less than the other two cream ads.

As seen in this research, these advertisements are coded according to the consumer information processing model.

3 Analysis & Conclusion

The role of advertising in consumer information processing phase is significant. Because the advertising acts as a guide for the consumer, to realize their consumer behavior after a situation or problem for the consumer. In this study, it is seen how this process works in herbal cream advertisements prepared for joint diseases. The related advertisements define the consumer and his problems, in the process of searching for information, oriented specifically for the consumer (those who complain about joint diseases, mostly elder people) and the problem and codify itself by aiming to stand out among the other alternatives. In addition, it is important for consumer evaluation to try to give confidence to the consumer with the return commitment after the product purchase. In the study, there are some common characteristics of herbal cream advertisements for joint diseases that are examined as an example. They all are aimed especially for the elderly; they declare that buy 1, get 1 for free; the product shipping is free and the price of all is 59 TL (Turkish Lira).

All three commercials draw attention to development of alternative eastern medicine. Therefore, "miracle" is also an important metaphor and product name in advertisements. In addition to that, old age, physical disability, neighborhood relations are also important rhetorical strengthening elements. Because these facts are common things of those who recommend each other according to the people who prepared these advertisements.

Religion is also used as an important common thing in many herbal product advertisements, even though it does not stand out much in these advertisements. The religion, which is one of the important medium of propaganda, is being used metaphorically through Islamic play sets and instruments, the benefits that the painkillers or preventive creams provide to pray in these advertisements. There are still ads that use religion as an argument. Some of the ads of such products that take religious motifs such as "one piece prayer dress (with Nal-i Sharif ring gift promise)" and "halal certified honey" broadcasted on some channels (in the end of 2017), attract attention. According to Elden, Ulukök, and Yeygel (2015), the main benefit of the product in cream advertisements is the ability to create a soft, silky touch. Color is also important in terms of corporation and brand image in such cream advertisements.

The use of celebrity names such as Latif Doğan and Songül Karlı in herbal cream advertisements examined in the study is also to corroborate the credibility and rhetoric as in the case of advertisements involving elderly people complaining about joint diseases. Because the society cares, trusts, and takes a role model in the context of the opinion leader of artists. In one of the advertisements, the presenter suggests in case of dissatisfaction "Bring it back if you don't like it, bring it back to Latif Doğan, and get your money." Also, looking at the advertisements with a professional look, these creams whose contents are similar appear to have entered the market by changing their name and the main ingredient in the cream is usually "black sesame."

Prior to 2018, "Glazed Herb Paste" ads were observed in some private TV channels where herbal cream advertisements were examined in this study. The arrival of winter months is emphasized in these advertisements which show that glazed plants are effective in preventing many diseases like herbal creams studied in this study. The benefits of the cream related to the advertisements are as follows: "Balances blood pressure, helps children to gain appetite, balances sugar, eliminates gastritis, lowers cholesterol, gives energy, relieves fatigue, relieves fatigue, is useful for impotence, provides weight control, increases strength, protects against cancer, adds beauty to the skin." Again in the related channels, a commercial called "Famous Pekin Cream" was observed in January 2018, and the cream marketed as buy 1 get 1 for free (59 TL [Turkish Lira] including shipping); it is alleged that the cream removes problems like pain, sores, joint rheumatism, even foot odor, sinusitis and migraine.

References

Advertising & Audiences, State of the Media. (2014). The Nielsen Company, Report.

Aktuğlu, K. I. (2006). Tüketicinin Bilgilendirilmesi Sürecinde Reklam Etiği. *Küresel İletişim, 2*.

Alikılınç, İ., & Özkan, B. (2018). Bir Sosyal Medya Pazarlama Trendi, Hatırlı Pazarlama ve Etkileyiciler: Instagram Fenomenleri Üzerine Bir Araştırma. *International Journal of Social Sciences, 1*(2), 43–57.

Altunışık, R., & Baş, Y. (2015). Tüketicilerin Ürün Satınalma Kararları Öncesi Aradıkları Bilgi Türlerinin Sınıflandırılması Üzerine Bir Araştırma. *Bilgi Ekonomisi ve Yönetimi, X*(I).

Avcı, U., & Avcı, M. (2004). Örgütlerde Bilginin Önemi ve Bilgi Yönetimi Süreci. *Mevzuaat, 7*(74).

Calvert, L. S. (2008). Children as consumers: Advertising and marketing. *The Future of Journalism, 18*(1), 205–234.
Chan, K. (2017). *Social issues of advertising*. Kowloon: City University of Hong Kong Press.
Collins, T. (2014). *100 ways to create a great Ad*. London: Laurence King Publishing Ltd.
Cömert, Y., & Durmaz, Y. (2006). Tüketicinin Tatmini İle Satın Alma Davranışlarını Etkileyen Faktörlere Bütünleşik Yaklaşım ve Adıyaman İlinde Bir Alan Çalışması. *Journal of Yasar University, 1*(4), 351–375.
Das, S. (2021). *Search engine optimization and marketing: A recipe for success in digital marketing* (1st ed.). CRC Press. https://doi.org/10.1201/9780429298509.
Dilek, S. (2016). Enformasyon ve Bilgiye Dayalı Yeni Ekonomi, *Kastamonu Üniversitesi İktisadi ve İdari Bilimler Fakültesi Dergisi,* Ocak 2016, Sayı: 11.
Duy, N. T., Mondal, S. R., Van, N. T. T., Dzung, P. T., Minh, D. X. H., & Das, S. (2020). A study on the role of web 4.0 and 5.0 in the sustainable tourism ecosystem of Ho Chi Minh City, Vietnam. *Sustainability, 12*(17), 7140. https://doi.org/10.3390/su12177140.
Elden, M., Ulukök, Ö., & Yeygel, S. (2015). *Şimdi Reklamlar*... (7th ed.). İletişim Publications 1101, Başvuru Dizisi 42, İstanbul.
Ellman, M., & Germano, F. (2008). What do the paper sell? A model of advertising and media bias. *The Economic Journal, 119*(537), 680–704.
Erbaydar, T. (2003). Halk Sağlığı Açısından Sağlık İletişimi. *C.U Journal of Faculty of Medicine, 25* (4), 45–51 (Special ed.).
Erdem, K. (2016). Tüketicilerin Demografik Özellikleri İle Satın Alma Tarzları Arasındaki İlişki: Aydın İlindeki Tüketiciler Üzerine Bir Araştırma (Adnan Menderes University Institute of Social Sciences Department of Business Administration Master's thesis), Aydın.
Ergül, H. (2013). Etkili Bir Sağlık İletişimi: Azerbaycan'da Bir Saha Araştırması. *H.U. Journal of Education, 28*(2), 166–180.
Fırat, A., & Azmak, E. (2007). Satınalma Karar Sürecinde Beyaz Eşya Kullanıcılarının Marka Bağlılığı. *Karamanoğlu Mehmetbey University Journal of Social and Economic Research, 2,* 251–264.
Frolova, S. (2014). *The role of advertising in promoting a product* (Thesis Centria University of Applied Sciences Degree Programme in Industrial Management).
Goldstein, G. D., Mcafee, R. P., & Suri, S. (2012, June 4–8). *Yahoo! research, improving the effectiveness of time-based display advertising*. EC'12, Valencia.
Greenwald, G. A., & Leavitt, C. (1984). Audience involvement in advertising: Four levels. *The Journal of Consumer Research, 11,* 581–592.

Heath, R., & Nairn, A. (2005, April). *Measure affective advertising: Implications of low attention processing on recall* (Working paper series). University of Bath, School of Management.

Hopkins, C. C. (2015). *Scientific advertising*. Createspace (First Printing).

Hülür, B. (2015). Yeni Medyada Sağlık Söylemi Üzerine Bir Analiz, *S B A R D, 13*(25) (Spring, 2015/1).

Hürriyet. (2017, August 24). Televizyonda gördü, 99 TL'ye kurban bıçağı diye aldı, limonu dahi kesemedi. http://www.hurriyet.com.tr/televizyonda-gordu-99-tlye-kurban-bicagi-diye-aldi-limonu-dahi-kesemedi-40560191.

Kara, T. (2017). Instagram'ın Sağlık İletişimi Bağlamında Kullanımı: Özel Hastaneler Üzerine Bir İnceleme. *Global Media Journal TR Edition, 7*(14), 7–22.

Karimova, Z. G. (2014). Defining advertising: A carnivalesque perspective. *International Journal of Journalismand Mass Communication, 1*(1), 002–010.

Karlsson, L. (2007 July 6). *Advertising theories and models—How well can these be transferred from text into reality?* Dissertation 10p, Strategisk Marknadsföring 51–60, Final Seminar, University of Halmstad.

Kasım, M. (2004). Ürün Reklam-Tanıtımı Açısından Gazetelerdeki Otomobil Reklamlarının İncelenmesi (1994 Yılı Hürriyet, Milliyet, Sabah Gazeteleri Örneği). *Selçuk İletişim, 3*(2), 137–144.

Keller, K. L., & Staelin, R. (1987). Effects of quality and quantity of information on decision effectiveness. *Journal of Consumer Research, 14*(2), 200–213.

Mackay, R. A. (2005). *The practice of advertising* (5th ed.) London: Routledge.

Mahon, N. (2011). *Basics Advertising 03: Ideation*. Lausanne: AVA Publishing SA.

Maya Tv. (2017). *Lazoğlu Sürmene Bıçağı* Reklam (Watched: September 2017).

Okay, A. (2014). *Sağlık İletişimi*. Derin Publications.

Özata, Z. F. (2011). Tüketiciyi Yönlendiren Güç: Öteki Tüketici. *IUYD'2011/2* (2).

Özcan, S. O. (2010). İnternet Pazarlama Faaliyetlerinde Tüketici Satın Alma Karar Süreci. *IUYD'2010/1* (2), 29–39.

Özdel, G. (2018, January). Yeni Reklam Ortami Olarak Vlog'larda Ürün Tanitimi Ve Anlam Üretimi. *e-Journal of New Media/ Yeni Medya Elektronik Dergi, 2*(1), 1–15. eJNM ISSN: 2548-0200, 2018.

Öztürk, G., & Öymen, G. (2014). Sağlık İletişiminde Sosyal Medya Kullanımının Stratejik Önemi: Türkiye'de Kalp Sağlığı ile İlgili Kâr Amacı Gütmeyen Kuruluşlar Üzerine Bir Değerlendirme. *Galatasaray University İleti-ş-im Journal*, 109–132.

Pektaş, H. (1987). REKLÂM NEDİR? İŞLEVİ VE ETKİLERİ NELERDİR? *On Dokuz Mayıs University Journal of Education Faculty, 2*, 221–230.

Phillips, B. J. (2006). REKLAMIN SAVUNUSUNDA TOPLUMSAL BİR PERSPEKTİF. Trans. F. Fulya Tepe, *Istanbul Ticaret University Journal of*

Social Sciences Year 5, No 9 Spring 2006/1 pp. 81–94. This article, originally titled "In Defense of Advertising: A Social Perspective", was published in *Journal of Business Ethics*, Volume 16, pages 109–118, 1997. Permission to publish the translation of the article was obtained from Springer by the translator of the article. Keywords and Turkish abstracts not included in the original text have been added to the article.

Pope, T. (2003). Making Sense of advertisements. In *Making sense of evidence-series on history matters: The US survey on the Web*. http://historymatters.gmu.edu.

Shachar, R., & Anand, N. B. (1998). The effectiveness and targeting of television advertising. *Journal of Economics & Management Strategy, 7*(3), 363–396, MIT.

Shafiq Gul, M., Ali Jan, F., Baksh Baloch, Q., Faheem Jan, M., & Farooq Jan, M. (2010). Brand image and brand loyalty. *Abasyn Journal of Social Sciences, 3*(1), 55–74.

Terkan, R. (2014). Importance of creative advertising and marketing according to university students' perspective. *International Review of Management and Marketing, 4*(3), 239–246.

Vakratsas, D., & Ambler, T. (1999). How advertising works: What do we really know? *Journal of Marketing, 63*, 26–43.

Van, N. T. T., Vrana, V., Duy, N. T., Minh, D. X. H., Dzung, P. T., Mondal, S. R., & Das, S. (2020). The role of human–machine interactive devices for post-COVID-19 innovative sustainable tourism in Ho Chi Minh City, Vietnam. *Sustainability, 12*(22), 9523. https://doi.org/10.3390/su12229523.

Yıldırım, Y. (2016). Tüketicinin Karar Alma Sürecinde Bilgi Kaynakları ve Güveirlikleri: Referans Grubu Olarak Yakın Çevrenin Etkisinin İncelenmesi. *Journal of Academic Approaches, 7*(1), 214–231.

Yılmaz, R., & Taşkıran, Ö. N. (2015). A model for mind-device dialectic and the future of advertising in social media age. In *Handbook of research on effective advertising strategies in the social media age* (pp. 1–16). Hershey, PA: IGI Global.

Williamson, J. (2002). *Deconding adverstisements ideology and meaning in advertising*. London and New York: Marion Boyars.

Additional Readings

Arvidsson, A. (2006). *Brands: Meaning and value in media culture* (1st ed.). New York: Routledge.

Bozgeyik, A. (2013). *Satış ve Pazarlama Öyküleri* (5th ed.). Akis.

Cialdini, R. B. (2017). *İknanin Psikolojisi: Teori ve Pratik Bir Arada* (Y. Fletcher, Trans.). Mediacat Books.

Das, S. (2020). Innovations in digital banking service brand equity and millennial consumerism. *Digital transformation and innovative services for business and learning* (pp. 62–79). IGI Global. https://doi.org/10.4018/978-1-7998-5175-2.ch004.

Das, S., & Mondal, S. (2016). *Yours digitally: Patanjali*. Available at SSRN 2737299.

Das, S., Mondal, S. R., Sahoo, K. K., Nayyar, A., & Musunuru, K. (2018). Study on impact of socioeconomic make up of Facebook users on purchasing behavior. *Revista Espacios, 39*, 28–42. Retrieved from http://www.revistaespacios.com/a18v39n33/18393328.html.

Das, S., & Nayyar, A. (2019). Innovative ideas to manage urban traffic congestion in cognitive cities. In *Driving the development, management, and sustainability of cognitive cities* (pp. 139–162). IGI Global. https://doi.org/10.4018/978-1-5225-8085-0.ch006.

Das, S., & Nayyar, A. (2020). Effect of consumer green behavior perspective on green unwavering across various retail configurations. In *Green marketing as a positive driver toward business sustainability* (pp. 96–124). IGI Global. https://doi.org/10.4018/978-1-5225-9558-8.ch005.

Das, S., Nayyar, A., & Singh, I. (2019). An assessment of forerunners for customer loyalty in the selected financial sector by SEM approach toward their effect on business. *Data Technologies and Applications, 53*(4), 546–561. https://doi.org/10.1108/DTA-04-2019-0059.

Elden, M. (2016). *Reklam Yazarlığı* (8th ed.). İletişim Publications 922, Başvuru Dizisi 39, İstanbul.

Gözler, K. (1999). *Anayasa Hukukunun Metodolojisi* (2nd ed.). Bursa: Ekin Bookstore Publications.

Güler İplıkçı, H. (2015). *Reklamlarda Tüketiciyi İkna Etmek İçin Kullanılan Stratejiler ve Reklam Örnekleri, Sosyal ve Beşeri Bilimleri, 7*(1), ISSN: 1309 -8039 (Online).

Gupta, D. K., Jena, D., Samantaray, A. K., & Das, S. (2019). HRD climate in selected public sector banks in India. *Revista ESPACIOS, 40*(11), 14.

Jain, S., Jain, V., & Das, S. (2018). Relationship analysis between emotional intelligence and service quality with special evidences from Indian banking sector. *Revista ESPACIOS, 39*(33), 3.

Karabaş, S. (2013). Pazarlama İletişim Aracı Olarak Tüketicilerin Reklam Denetimine Karşı Tutumları: Reklam Etiği. *Journal of Çankırı Karatekin University Faculty of Economics and Administrative Sciences, 3*(1), 143–157.

Kwak, H., Changhyun, L., Hosung, P., & Moon, S. (2010). What is Twitter, a social network or a news media? *Proceedings of the 19th International Conference on World Wide Web.*

Mestçi, A. (2017). *İnternette Reklamcılık: Dünyada ve Türkiye'de İnternet ve İnternet Reklamcılığı Kavramları* (2nd ed.). Pusula Yayıncılık ve İletişim.

Michaels, V. (2014). *Reklam Aşkı* (1st ed., M. Sağlam, Trans.). Nemesis.

Mohanty, P. C., Dash, M., Dash, M., & Das, S. (2019). A study on factors influencing training effectiveness. *Revista Espacios, 40,* 7–15. Retrieved from http://www.revistaespacios.com/a19v40n02/19400207.html.

Mondal, S. R. (2020). A systematic study for digital innovation in management education: An integrated approach towards problem-based learning in Vietnam. In *Digital innovations for customer engagement, management, and organizational improvement* (pp. 104–120). IGI Global. https://doi.org/10.4018/978-1-7998-5171-4.ch006.

Mondal, S., Das, S., Musunuru, K., & Dash, M. (2017). Study on the factors affecting customer purchase activity in retail stores by confirmatory factor analysis. *Revista Espacios, 38*(61).

Mondal, S., Mall, M., Mishra, U. S., & Sahoo, K. (2017). Investigating the factors affecting customer purchase activity in retail stores. *Revista Espacios, 38*(57), 22.

Mondal, S., & Sahoo, K. K. (2020). A study of green building prospects on sustainable management decision making. In *Green building management and smart automation* (pp. 220–234). IGI Global. https://doi.org/10.4018/978-1-5225-9754-4.ch011.

Nadanyiova, M., & Das, S. (2020). Millennials as a target segment of socially responsible communication within the business strategy. *Littera Scripta, 13*(1), 119–134. https://doi.org/10.36708/Littera_Scripta2020/1/8.

Reeves, R. (1961). *Reality in advertising*. New York: Alfred A. Knopf.

Sharma, E., & Das, S. (2020). Measuring impact of Indian ports on environment and effectiveness of remedial measures towards environmental pollution. *International Journal of Environment and Waste Management, 25*(3), 356–380.

Singh, I., Nayyar, A., & Das, S. (2019). A study of antecedents of customer loyalty in banking & insurance sector and their impact on business performance. *Revista ESPACIOS, 40*(06),11.

Singh, I., Nayyar, A., Le, D. H., & Das, S. (2019). A conceptual analysis of internet banking users' perceptions. An Indian perceptive. *Revista ESPACIOS, 40*(14), 1.

Singh, L. B., Mondal, S. R., & Das, S. (2020). Human resource practices & their observed significance for Indian SMEs. *Revista ESPACIOS, 41*(07). Retrieved from http://www.revistaespacios.com/a20v41n07/20410715.html.

Singh, S., & Das, S. (2018). Impact of post-merger and acquisition activities on the financial performance of banks: A study of Indian private sector and public sector banks. *Revista Espacios Magazine, 39*(26), 25.

Singh, S., Mondal, S., Singh, L. B., Sahoo, K. K., & Das, S. (2020). An empirical evidence study of consumer perception and socioeconomic profiles for digital stores in Vietnam. *Sustainability, 12*(5), 1716.

Topsümer, F., & Elden, M. (2016). Reklamcılık Kavramlar, Kararlar, Kurumlar (15th ed.). İletişim – 763, Başvuru – 18.

CHAPTER 4

Games Features for Health Disciplines for Patient Learning as Entertainment

Shariq Aziz Butt, Saikat Gochhait, Shama Andleeb, and Muhammad Adeel

1 Introduction

Serious games are used mostly for gaming and health combinations. Serious games are defined as games which are designed with the objective of health care rather than entertainment. Sawyer describes health care as a field of enthusiasm for games due to his extensive study in this domain of research. He highlighted so many major points that should be kept in mind to continue research in this domain. Sawyer's points have helped in the design and development of games for health care. There are many

S. A. Butt (✉)
The University of Lahore, Lahore, Pakistan

S. Gochhait
Symbiosis Institute of Digital and Telecom Management, Symbiosis International University, Pune, India
e-mail: saikat.gochhait@sidtm.edu.in

S. Andleeb · M. Adeel
The Superior University, Lahore, Pakistan

S. Das and S. Gochhait (eds.), *Digital Entertainment*,
https://doi.org/10.1007/978-981-15-9724-4_4

games designed and developed for activities like physical wellness, clinical experts, and patient reminders. Each game is developed to give benefits to the developer and the patient for assessing the game's results in terms of the game's support, and to help to patient in his/her disease monitoring (Leung, 2014; Ohn, Ohn, Souza, Yusof, & Ariffin, 2019).

Gaming for health care is the domain in which describe the influence of games on health (Harviainen, 2019; Veselka, Wijesingha, Leatherdale, Turner, & Elton-Marshall, 2018; Zendle & Scholten, 2020). Through organized and semistructured prompts, we gathered high reactions to explicit gaming and gaming's ideas from those specialists (Ricciardi & De Paolis, 2014; Yap, Yap, & Yap, 2015). We used their responses to describe the attributes required for the game's design for health care (Cooper & Edgett, 2012; Jackson et al., 2020; Tan et al., 2020).

2 Health Consequences of Gaming for Health

A very few number of games were found to support and enhance solid ways of life and to build illness information for the benefit of patients and individuals with diabetes. Games likewise have indicated beneficial outcomes on mental, physical, and social health in more seasoned grown-ups (Aksoy, 2019; Sobkow, 2017). Furthermore, numerous games for health were designed and developed but have no positive benefits either for patients or stakeholders. Therefore, in games for health, clinical investigations ought not to simply discover that a specific game is successful, but should assist analysts in understanding why it is powerful (Butt et al. 2016). This cannot remain managed without a primary knowledge of the basic game's characteristics (Bosworth, 2012; Gafni, Achituv, Eidelman, & Chatsky, 2018).

3 Gaming and Health Frameworks

The research community has analyzed the existing work on gaming for health as conceptual frameworks. A proficient framework for games for health and their scientific categorization can be used to design and developing games by their features (individual, proficient practice, general health engaged), and by well-being activities (helpful, instructive, or informatic). In the study (Arnab et al., 2015) an additional edge showed helpful games as per three fundamental foci of player games, and the treatment and connections among them. The systems provide significant basic experiences, for example, distinguishing regions which have unexplored potential for games for health.

Different frameworks were proposed by the research communities, but all have some limitations in terms of factors that need attention in designing games for health care.

As a framework proposal relating to the design and development of games for health conditions, the game designers design a game's prototype with some features of health for patient learning during game-playing as described in Fig. 1 (Olszewski & Wolbrink, 2017). Then the game was developed and used for testing with on patients. It takes several attempts to generate better results on games for learning. In the results, the game's framework has limitations in terms of usability and testing of games.

Another framework proposal contains the concept of design of a game for health, sets some standards for game design for a specific group of patients, then develops the game and obtains some results. It also carries out some rounds with the patient to evaluate the effectiveness of the

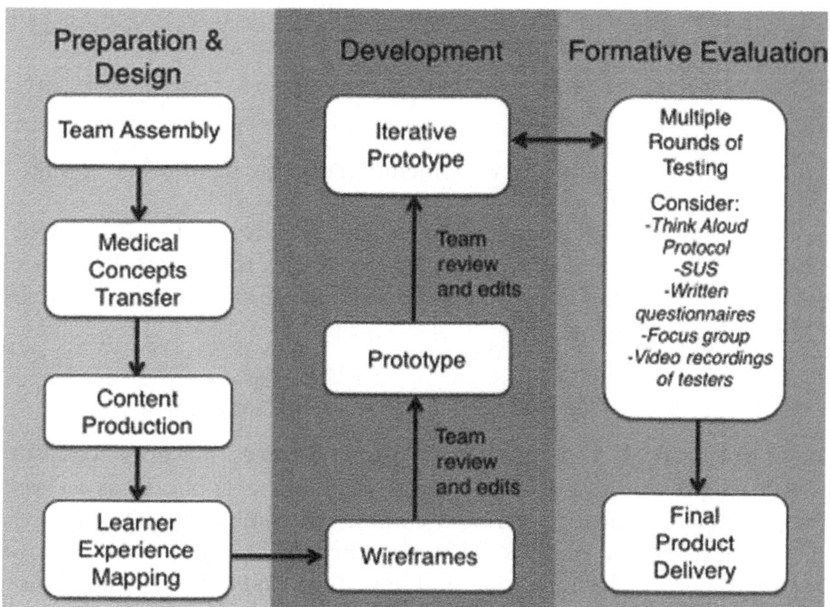

Fig. 1 Framework for health gamification (Olszewski & Wolbrink, 2017)

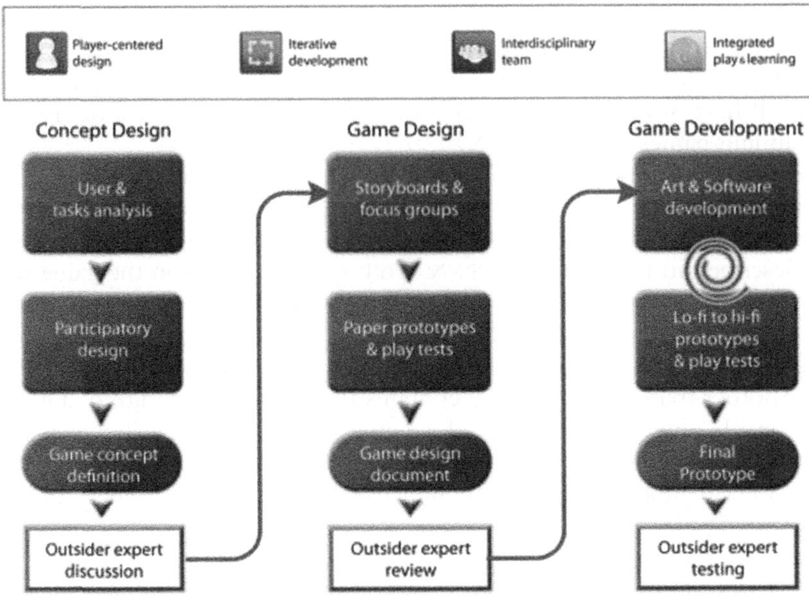

Fig. 2 A standardized framework for health gamification (Abeele et al., 2011)

framework for the game development for health mentioned in Fig. 2 (Abeele et al., 2011).

These frameworks have limitations for the targeted audience, i.e., patients with their disease type. This study reveals some factors that need to be addressed in the designing and development of games for the health and learning of patients.

4 Behavior Change Theories

Many professionals have worked on the utilization of games for health as well as presenting some new ideas for gaming applications to well-being. They developed games and conducted studies on games after the playing of games by patients. They also checked changes in the behavior of patients while playing games and gave their responses. Some of these games for behavior change are smoking-counteraction, exercise-based therapies, and medicine reminders. Some recent studies have argued that the behavior change is the most important aspect when you play games for health learning (Kharrazi, Faiola, & Defazio, 2009; Kim, Prestopnik, & Biocca, 2014; Payne, Moxley, & MacDonald, 2015).

4.1 Using the Behavior Concept for Designing Gaming

A review of smartphone applications for health and general health noted that most applications do not including much health-conceptual information, while fitness-focused video games have been found that are useful for losing weight. There is also a need for studies that present the game's testing theories or models used for health care (Starks, 2014). Due to the smaller amount of work done in this area, the design of games for health is facing some limitations and not is producing proficient results. Furthermore, there is also a difference between evaluating these findings and recognizing how to better integrate designing processes in the creation of games for fitness or health assessment (Isbister & Mueller, 2015). In the design enhancement of players' physical activities and social support, the research built a model of mediating factors that must be considered to enable the field to better assess game design and enhance its design and development in the future. Subjects from Lieberman's mix shows in this course of action rules such as proposing giving players depictions that help with efficient new developments, which interfaces to the chance of self-sufficiency. These medicating components have not been adequately connected with unequivocal games' features or design choices (Mueller, Khot, Gerling, & Mandryk, 2016; Segura & Isbister, 2015).

The gap between game design and the concept of behavior is reflected in methodologies for gaming for health architecture. The five-point methodology for health and medicine games in Playmatics stresses the researchers and game designers might well have separate areas of expertise within the same team. Schell Games' process is assisted by the Framework for Game design providing seminars and opportunities for debate to assist stakeholders to work together. However, it is a framework with wide-ranging methods to transform games and but does not focus on health directly (Harpstead, 2017).

4.2 Game Mechanics and Features

Research is focused on suitable ways to deal with the best structure for the prosperity of games that consolidate the development of game mechanics, and area specialists suggest additional ideas that need to focus on these kinds of features in order to create successful games for well-being. In any scenario, games must make orderliness a premium to enable players in the direction of perfect thriving and knowledge results; thoughtful beneficial game mechanics is basic for effort (Mosca, 2017). Additionally, various games mechanics can facilitate or prevent desired health outcomes in different forms (e.g., to promote or restrict physical activity), and it is vital to consider the interaction of games and health (Ricciardi & De Paolis, 2014). Much of a game's elements, including multiplayers, in the sense of design and development, are actually not well understood (Frome & Martin, 2019).

4.3 *Standards for Games for Health*

These opinions concerning games for health can concentrate on the health situation addressed by the game "Running Zombies!" For instance, an inclusive, non-judgmental definition of exercise is reflected, whereas certain games are far more restrictive for obesity. Prosperity games can comparably characterize perceptions about how prosperity directs change-capacities by consolidating portions of prosperity and so lead to change structures. Thus to recommend a careful framework for seeing the credits that one needs to give in a game. For the considerations, the thriving game "Lit2Quit" looked out for tobacco smoking; taking everything into account, its position is that gaming ought to support players' previous need to stop, instead of trying to convince them to stop. This position originates from esteem-responsibilities with respect to game designers; for example, regarding the self-rule of smokers who better comprehend their own life conditions; these qualities are thusly reflected in explicit structure selections. These include, diminishing the desire to smoke in a procedure started by the player. A worthwhile intelligent procedure is particularly significant when the developers associated with a game's creation originate from various fields of expertise, with no guarantees, frequently the case in games for health, and seeing how these various team members distinguish and interpret features and qualities that are essential to collaborating in games in progressively viable ways (Calderón, Trinidad, Ruiz, & O'Connor, 2018; Gauthier et al., 2019; Göbel, Vogt, & Konrad, 2018).

5 Methods

Our study starts by examining the current prospect of well-being gaming, and collecting types of games for health. It supports this study to develop a progressively significant appreciation of existing well-being games in research and business settings and provides a list that can help to choose which games to play. We have selected some games from health-games catalogues, but there is no existing collection for well-being gaming adhered to the standards depicted underneath or totally comprehensive with respect to gaming (Butt, 2016).

5.1 *Health Gaming Number*

Our gaming-for-health data contains 320 successful gamings from a plan of base, as well as past appearances, earlier research databanks (Verschueren, Buffel, & Vander Stichele, 2019), and Google's Android application collection. In addition, we have Apple's application collection (watchwords: "games for flourishing"; "games for prospering the board"; "tribulation the experts games"). We proposed to adhere to the developers and designers' needs, so that games that were exceptional in themselves were joined, and applications that were insignificant "gamification" were exculpated. Games proposed for preparing clinical authorities were moreover avoided. To help to get together, games were mandatory in having an inside circle (Kayali, Luckner, Purgathofer, Spiel, & Fitzpatrick, 2018), or had stages of examination in which a player practiced the influence of the game and gave feedback.

We picked sixteen games used in all ace assessment parties (see Table 1). We picked an optional subset of playable games and kept an eye out for them for replacements, until we had a changing delegated starter of the more obvious quantity for sensible investigation-based and business games, in addition to gaming stages.

5.2 *Training Method*

The study arranged examination meetings with eighteen members (P) (7M, 11F): game plan specialists ($n = 7$), conduct well-being specialists ($n = 5$), and health game specialists ($n = 6$). Input specialists more likely than not finished, or then again sent at least one game's well-being intervention. This study likewise expected to incorporate specialists from

Table 1 Games used for study

Gaming	*Health Scenarios*
Cigma	Cancer
Ashma Dodge	Asthma
DodgeBall	Exercises activity
Chicken Farm	Nutrition
Colorfall	Mental health, physical activity
Dex: Virtual Pet	Diabetes
Elude	Depression
Escape from Diab	Diabetes
MySugr Junior	Diabetes
Mindless eating challenge	Nutrition
Monster Manor	Diabetes
Pack and Marlon	Diabetes
Rex Ranon	Smoking cessation
Snow world	Pain management
Sortify	Nutrition
Virtual Iraq	PTSD

various foundations (e.g., two analysts; what is more, individuals from industry). The selected games are mentioned in Table 1.

We enlisted by observing subjects' conversations and by snowball testing, and by direct investigation meetings both face to face and remotely. All participants gave cultivated assent and every single remote member marked for acceptance. This paper's material for participant meetings comprised of physical game cards, although remote meetings utilized advanced cards. All meetings were sound- and video-recorded and kept going for roughly sixty minutes. Study meetings were comprised of three stages: an exploratory stage where specialists portrayed an obscure game so anyone might hear, dependent on pictures; a trading stage where the participant looked at gatherings of three games; and an closed card sort that ordered fifteen games. These three stages, picked to accumulate inputs in numerous ways, both organized and open-ended, are described next.

5.2.1 Stage I: Exploration

Members previously viewed pictures from a well-being game (Cigma) and asked what type of data would be required by participants to comprehend the game. We utilized verbally processed conventions to inspire the members' point of view on what they comprehended from the photographs, what data they had a chance to see, and what questions they would raise in order to learn the game for participants' best capacity.

5.2.2 *Stage II: Open Card Sort*

In the second stage, members directed an open-finished card sort for every one of the fifteen well-being games from the trading stage. We requested that every member sort the games any way they believed best, giving no additional guidelines or limit points on how numerous groups ought to be made or of what kind. When the card sort was finished, members clarified their arrangement based on the scientists.

5.3 *Data Examination*

We broke down open-card-sort information via illustrative measurements and similarity networks to decide normal game pairings. In translating spoken reactions from members, three scientists led an iterative, inductive topical examination of the group of three correlation round outcomes, by utilizing a steady relative methodology (Mehta & Pandit, 2018). Numerous ideas were coded inside every round. Through conversation among the exploration group, we utilized these ideas to infer seven more elevated level subjects.

5.4 *Discoveries and Discussion*

Out of 270 rounds, we found seventy-six examinations referencing a set of different events by specialists as mentioned in Table 2. These seventy-six ideas followed the seven logical subject levels like **player quality**: the player quality beyond of knowledge and practice in gaming. **Well-being modeling**: the most important feature in the game should be how the well-being concept is implemented, presented, and user friendly with respect to game design. **Gaming world**: this portrayed practical parts of the game, for example, the story. **Game mechanics**: gaming measures and interrelated components, for example, the **designing of stages**: it leads to player for enagage with game playing and enhance player's mentality to learn. **Settings**: or on the other hand, the way that predictable correspondence interfaces with this current reality; and **Results**: how players or their general common factors are different from one player to another for understanding.

Table 2 Factors for gaming for health

Player features in playing games	*Health Framework*	*Gaming world*	*Player behavior*	*Context*	*Outomes*
Cultural assumptions	Health conditions	Game user friendly	Interest to play	Targeted audience	Behavioral change
Gender	Health issues	Game features	Interest after the game play	Timing of game to play	Learning
Patient health conditions	Type of disease	Game easy to play	His/her interaction while playing the game	Context of use	Fun with effectiveness
Player expertise and level of mentality for gaming	Disease seriousness	Game designing	His/her decision after playing the game	Game interaction	Evaluation
Patient leaning through gaming	Health framework accuracy for targeted game	Game catalogues	His/her satisfaction after play	Playing repetition	Health improvements

In this section, we start surveying the tendencies where subjects were seen especially by expert get-togethers, starting with a short design before jumping into unambiguous outcomes. We likewise offer further evaluations between get-togethers, including associations by thought and by games.

6 Game Design Experts

While game structure specialists referenced well-being ideas, they were clearly sensitive to the gaming aspects of games for health, tending to utilize language that referenced different games as opposed to other well-being meetings. They additionally habitually thought about the player as an individual with self-rule, as opposed to concentrating exclusively on their well-being condition. At long last, game specialists based their conversation for the connection between gaming and well-being subjects on the reconciliation of gaming with well-being conduct. Our studies examined these distinctions with greater clarity.

7 Gaming Parts of Games for Health

In games for health, game design specialists considered the players' well-being objectives (a player trademark) and the well-being conditions anticipated throughout the game (some portion of the well-being prototype). At the same time, "Packy and Marlon" and "Rex Ronan" seem to be really comparative as to their levels, and in any event they seem to have a few similitudes. These specialists found that game groups of three could be portrayed in different notable manners, frequently with one well-being engaged and one game-centered heading. As the P10's reaction appears, the game and well-being topics in some cases proposed diverse game pairings.

8 Patient Experiences as Gamers

Specialists in collecting were especially touchy to play the attributes, surrounding the player as an individual with self-sufficiency and seeing earlier educational encounters. P8 talked about the significance of players' experience when separating Escape from Diab, clarifying that "it appears to be founded on a explicit arrangement of rules. However 50% of them cannot use the controls at all and go through thirty minutes simply looking at a separator. In essence, you cannot get the experience out of the game in the event that you try not to carry the capacity to explore the controls preceding the experience."

One game player showed an encounter with regards to sexual orientation and culture, the main participant in any gathering to notice either idea. The game participant who talked about sex did as such during trading, taking note that "this could simply be a poor idea, yet I feel like these two (Mindless Eating Challenge and Asthma Dodge the Bronchiasaurus) could be more geared towards young ladies, or possibly to a sexually unbiased sort of crowd. I do not have a clue whether this is only the absence of thought about style, yet (Rex Ronan) certainly doesn't shout 'young ladies'" (P18). A couple of rounds later, the equivalent participant addressed social suppositions in the setting of game style and earlier information, clarifying that "there's very social suspicion (in Monster Manor) about comprehending what option Frankenstein is having to decipher this stylish and ghoulish sort of thing" (P18). While numerous specialists in all gatherings examined the crowd in terms old enough (youngsters versus grown-ups) or well-being status (analyzed with a condition versus the all-inclusive community), no other participant addressed the significance of sociocultural inspection in the structure of well-being games. The truth be told, we expect to experience this all the more every now and again, given that the scholarly examination of games frequently receive explanatory methodologies that treat games as social writings and relics in a sociocultural setting.

9 Incorporation of the Gaming Process and Fitness Performance

P14 furthermore noticed that after these observations were expressed designers would roll out these improvements that might be reflected, taking everything into account. Then, other game structure specialists communicated concerns about an absence of joining among game methods and well-being ideas in certain games. P18, for example, did not think Ashma dodged the exhibited satisfactory incorporation: "I can't envision this could attract a similarity to any true conduct in kids, since it resembles your little creatures going through the wilderness." P2 felt similarly about Sigma, clarifying that "it seems like they're acquiring a lot of mechanics from different games, without an away from those of why they've gotten those mechanics." For the gaming participants, game prosperity fuse was a normal extent of relationship among games, and they believed such joining to be a marker to make gaming be low in efficaciousness as a mediation. A strong development concept is required for games development (Butt, 2016).

9.1 *Health Specialist with Ideas and Experience*

By and large, social well-being specialists were destined to reference ideas inside the significant level topic of well-being models. Like game specialists, conduct well-being specialists were too worried about player attributes, particularly independence of an alternate kind; they concentrated on player self-governance according to one's own well-being and corresponding to one's readiness to make a move and modification conduct. The specialists additionally separated "fun" in games as unmistakable from getting the hang of observing, or other beneficial well-being exercises. At last, their valuation of the relationship between game and success content in lodging game mechanics could be seen as a portrayal of the thriving model and substance, as opposed to two relatively raised level contemplations.

9.2 Patient as Player for Examining Their Health

Much of the gathering's conversation was about well-being ideas created about player self-rule comparable to their well-being. Social well-being specialists most generally looked at game trios dependent on the highlights of well-being objectives, well-being conditions, and well-being topics; yet they did as such with regard to the results, taking note of whether and how conduct modification steps and readiness for the activity were measured in the plan.

Strikingly, a couple of experts focused on the essential fact that playing a game might be intruding on player independence, else if nothing else could adversely influence the player's real-life prosperity experience: analyzing Cigma in Stage I, P11 felt that, "I would lean toward not to express the game is making light of what they're encountering, anyway you are starting now to defy harmful development, malady is your life when you're experiencing chemo. Would you truly like to play a game with a malignant growth?" Conduct well-being specialists additionally differentiated games by referencing parts of the setting, particularly the certifiable associations. P15 clarified that Monster Manor was unique in relation to Escape and Rex Ronan because of "your individual conduct, you have to follow that all together for the ongoing interaction to be fascinating," while P13 felt that "Beast Manor and Packy and Marlon are comparable on the grounds that you're logging your own data." Other well-being specialists talked about the intelligence and the certifiable setting of both Mindless Eating Challenge and Dex: Virtual Pet. Furthermore, interestingly to Fix, Frank P3 noticed that "both Monster Manor and DodgeBall appear to be more grounded in what individuals' well-being conduct is in reality." Conversation on genuine world settings and linked topics displayed an emphasis on the player as opposed to gaming obvious from social well-being specialists, however, such that despite everything identified with player conduct, there was also ongoing interaction. In Phase I of the investigation, social well-being specialists additionally would in general spotlight on how a player's true well-being might be influenced by the ongoing interaction. P12 accepted that Cigma's objective was to show players how chemotherapy functions; furthermore, in what way it may influence them, while P15 was interested to study the educational program substance of gaming and in what way intently it cut to the real world, stressing that a real reproduction of chemotherapy may be stunningly distressing. Strikingly, although numerous members discussed the Cigma general crowd, just a single member (P11) thought about whether there was an increasingly explicit objective crowd than only "kids with disease."

9.3 Gaming for Fun and Well-Being

Leading well-being specialists regularly examine the degree of "reality" of the game. In our findings from the participants, P12 characterized almost all games in the health surroundings. He also grouped the three games as gaming for fun, not specifically for health. The participant P13 also responded games as fun after playing all games such as Snow World, Rex Ronan, Colorfall, and Sortify. He grouped these games as differentiated from education games, and well-being gaming. There is a defined categorization of fun gaming, but the participant P15 mentioned that "I prefer the fun of the game as compared to learning of game features." Cigma is a game for fun more or less than learning. The remaining participants stated that MySugr Junior is the only game to play (P12), and P11 stated that Colorfall and Snow World are some types of meaningless games. They mentioned that fun does not necessarily mean easy for people. It is an idea that game developers need to keep in mind while designing the games. This concept elaborates the clear distinction about gaming for fun and gaming for health. The game developers should take the special concern of health experts to set some characteristics for games specially designed for health. The health games should be more precisely for health, not for fun. There should be clear differences between the fun and health features of gaming.

10 Context with Interface

Games for health specialists, as conducted well-being specialists, were well on the way to make trading examinations dependent on the well-being objectives and well-being conditions. Games for health specialists were considerably additionally concentrated on the setting of utilization encompassing a game. P16 differentiated Beast Manor from other games by clarifying that it "stands isolated pretty unmistakably, considering the way that it's an adaptable game that is needed to be joined into a diabetic's life as they track their diabetes data, anyway the other two are a substitute occasion for the remainder of the player's life." Similarly, P15 found that "Dex is the weirdo since it should be utilized over some dubious schedule opening and you're getting every day rewards, while Virtual Iraq, you would do possibly once in a sitting and a brief timeframe later conversation about it, moreover, Ashma Dodge the Bronchiasaurus is equivalent, you could essentially play it once and get the movement, it won't reward you for reestablishing the way that Dex does." For some games for health aces, a discussion of the flourishing games was clearly not complete without this pulled in discussion of the wrapping setting. Moreover, this thought of setting was at times limited considerably more unequivocally to the degree of the social setting by games for health specialists. Strangely, this was a rule get-together to demonstrate the social relationship inside Cigma in Phase I.

11 Results and Aims

Game of health pros in our examination furthermore deliberated the results of both inner and outer aspects of gaming. P5 observing that "kid-friendly games have fundamental objectives, but the objectives are less clear for Virtual Iraq." P16 clarified gaming purposes as well, strengthening that Packy and Marlon and Chicken Farm "seem to have constrained game play, a target that you can reach, however, Monster Manor, considering the way that its normal for lead change and because of the social mobile DNA there is impressively increasingly boundless. The objectives are progressively powerful regarding, I need to make it to the following level, or I need to locate this specific beast thing like that. At that point, once one objective is finished, more objectives become evident, while the other two resemble the sorts of games that have a closure, where you arrive at your objective and you are finished." P1 thought about an extra set of three by saying that "these two Dex and Packy and Marlon are indistinguishable in light of their substance, that they're in diabetes training, it appears like it's here to incite them to have sound propensities and to instruct them about their illness, though Elude is not truly instructing you about discouragement." Different specialists referenced that particular games have a target of demonstrating the player guidelines to manage themselves or others, taking everything into account (P5) or an expectation community in this circumstance, evasion of diabetes (P7). Games for health specialists, also observed the chance of negative gaming results having a latent negative or demotivating influence on the player's true thriving, and they were the standard ace group in Phase I to address potential worries of the game for the normal interest gathering. P1 considered games to be used as a bracing instrument for kids experiencing chemotherapy; in any case, she saw the criticalness of ensuring the gaming was not excessively hard for the target, so that it didn't have this demotivating influence on the player. P5 resonated with this worry, and called Cigma a "somewhat delicate game."

11.1 Evaluations Between Professional's Groups

Inside Phase 2 of the examination, six through fifteen inquired specialists had to think about game sets of three dependent on one of the ten allowed ideas. By differing games being thought about for a guaranteed idea, we watched normal topics rise. In the play techniques round, for example, game specialists were destined to talk about game mechanics, particularly win/misfortune conditions, separating games by clarifying that "some games are letting you play through and you can lose, but here you can pass on yet you can't lose the game" (P9). Well-being specialists, then again, were bound to portray play procedures and game achievement as far as the game world idea of remunerations, noticing that for certain games "you could come in with your insight and play to your quality to get further focus in the game" (P11). However, well-being specialists appear to comprehend awards as the fundamental spark for play regardless of game results. There were further gathering contrasts in the subjects referenced in the crowd round. Game plan specialists discussed game stylish intuiting, for instance, that a gaming was equipped in the direction of kids "just from the interfaces and hues and their selection of designs" (P10). Games for health specialists, notwithstanding, additional frequently examined well-being ailment and well-being status, separating the crowd between "kids that most likely have been determined to have something with a ceaseless sickness, while this other game is only for anybody" (P1). Finally, well-being specialists every now and again examined crowds in the setting of well-being topics and inspiration, with P15 clarifying that a few games "appear to be focused on individuals who really are hoping to make a change in their diet or physical action" (P15). Some doled out ideas that incited comparative subjects over gathering.

Table 3 Illustration of expert results

Expert Groups	*Important results*	*Corresponding high-level features*
Game designers	Gaming characteristics of GH4 Patient as user with some experience Combination gaming features with health	Gaming world Players' experience Gaming with health modeling
Behaviroual well-being specialists	Patient intention to his/her own health for "fun as results"	Patient features for health game modeling
GH4 specialists	Aims with some results Gaming features produce health results	Context Results

Game objectives, for example, disclosed to members as "in-game objective goals," still oftentimes prompted a conversation of well-being objectives for specialists in all gatherings. One game participant noticed that in Dex "the Tamagotchi position causes the objectives to adjust intimately with what the realized goals would be" (P17), while a well-being participant surrounded the correlation of game objectives by clarifying that "the general objective is to be more advantageous and make nourishment decisions, and improve sustenance" (P11). Settings of utilization additionally observed all gatherings every now and again talking about similar subjects, in particular well-being objectives and, furthermore, rehashed play. For these doled out ideas, the changes in reactions further emphasize bunch contrasts, yet likenesses offer a guarantee for interdisciplinary comprehension and correspondence. Table 3 illustrates the results and views of experts.

11.2 Comparisons by Game

We analyzed conversation and examination by game, both among and across participant gatherings. This is significant for various reasons. To start with, there are genuine contrasts between the games in our investigation and we need to explain the specific subjects each game may have been bound to inspire. In particular, however, since we are at last spurred by a craving to characterize and distinguish game highlights in ways that are helpful to the scope of partners, it is essential that we expect to comprehend singular games. This does not bring down the significance of looking at topics and includes participant gatherings; rather, it permits us an extra, progressively engaged perspective on the games that our specialists were looking at, and which started up the ideas investigated to date. Taking gender as the topical after effects of the open card sort of game, we saw that a few games in our corpus were more prone to evoke specific topics over every participant gathering: well-being objectives, flourishing conditions, and game requesting were the subjects most usually used to store eleven of the fifteen games. For both Monster Manor and Mindless Eating Challenge, in any case, genuine help was routinely an amazing piece of time utilized for the layout, and P15 collecting this got "real-world credit" and P3 showed these games as "a smidgen at time life" additionally, "remunerating dependent on attested direct." For Dex and for Virtual Pet redirection was the subject of decision, while in Snow World most bosses by and large contemplated its degree of "reality." Therefore furthermore, Colorfall was routinely portrayed by its nonattendance of thriving relatedness, with different experts saying that it "doesn't fit with any game plan" and "[I] can't make sense of what the flourishing target is" (P16), or that its a "non-success" (P7) or "fun game" (P14). This again serves to show a normal division between games that are surely authentic and what is logically "helpful" and different games that are verifiably not helpful. Such remarks depict the difficulties of unpretentiously sorting out authentic substance in games, and the balance between fun and guidance in making and consolidating certifiable prosperity games. We can in a like manner see game examples by considering similarities, and additionally, contrasts in how games were accumulated by individuals from every one of the three ace social events. During the open card sort, Fix Frank, DodgeBall Rubble, and Escape from Diab were accumulated by every part of the prosperity ace social occasion for instance, while each was gathered with the others for a far less significant part of the time by both game and games for health experts.

12 Games for Health: Outside the Summation of Its Components

Starting a discernment on the points of view of these diverse associated social events can help us to improve interdisciplinary correspondence. We surmise that different games for health do not sufficiently embody the success lead change theory, as the making appears, considering the way that the game similarly as games for health assistants are choosing different hands-on plan choices without the thriving point of view as a top need. By refining the discussions between accomplices during the structure framework, we can progress the game plan of games for health. For instance, games for health workshops could be worked with the existing courses of action instruments, for example, the games structure (Butt, 2016), yet deliberately sifted through with our seven subjects as a perfectly healthy interdisciplinary correspondence. Second, this work uncovered the considerations and standards of major games for health tendencies, as explicit from the individual controls, surely of activity and flourishing direct modification. The subjects and reactions of the games for health ace pack in our assessment were not just the entire range of the focus and reactions of game structure and direct flourishing specialists; for instance, games for health specialists were the essential ace get-together to decide the number of players or game capriciousness as subjects. Legitimately, for every accessory get-together, we have a quick overview of the reflections that they are in danger of using in confining games for health and what is more, an outline of the contemplations that they are in all likelihood going to discard in their official review. We can make arrangement "prompts" set up not just regarding our raised level subjects, in any case, on unequivocal thoughts, for example, recalling players' connected contributions (usually observed by simply game makers), concentrating on their status to change lead (simply prosperity participants), and seeing the social setting in just games for health.

13 Conclusion

Game design and development is a major domain of research in entertainment. The gamification of health conditions is the targeted domain of the research community. Health games are developed for the learning paths of patients with different diseases. The most important factor in health gamification is that games features should be appropriate to the disease for which the game has developed. In this study, we highlighted these factors and conducted a study using different patients as users to elaborate these factors. We also highlight some limitations of those frameworks that are not so far useful for game development for health. We also performed a detailed comparison on the existing games developed for health.

Acknowledgements This research work was supported by the University of Lahore, Lahore, Pakistan and the Symbiosis Institute of Digital and Telecom Management, Symbiosis International University, India.

References

Abeele, V. V., De Schutter, B., Geurts, L., Desmet, S., Wauters, J., Husson, J., ... Geerts, D. (2011, October). P-iii: A player-centered, iterative, interdisciplinary and integrated framework for serious game design and development. In *Joint conference on serious games* (pp. 82–86). Berlin, Heidelberg: Springer.

Aksoy, E. (2019). Comparing the effects on learning outcomes of tablet-based and virtual reality-based serious gaming modules for basic life support training: Randomized trial. *JMIR Serious Games, 7*(2), e13442.

Arnab, S., Lim, T., Carvalho, M. B., Bellotti, F., De Freitas, S., Louchart, S., ... De Gloria, A. (2015). Mapping learning and game mechanics for serious games analysis. *British Journal of Educational Technology, 46*(2), 391–411.

Bosworth, A. (2012). Keas: Developing a successful game-based employee wellness program. *Games for Health: Research, Development, and Clinical Applications, 1*(3), 189–191.

Butt, S. A. (2016). Analysis of unfair means cases in computer-based examination systems. *Pacific Science Review B: Humanities and Social Sciences, 2*(2), 75–79.

Butt, S. A., Abbas, S. A., & Ahsan, M. (2016). Software development life cycle & software quality measuring types. *Asian Journal of Mathematics and Computer Research*, 112–122.

Calderón, A., Trinidad, M., Ruiz, M., & O'Connor, R. V. (2018, September). Towards a standard to describe and classify serious games as learning resources for software project management. In *European conference on software process improvement* (pp. 229–239). Cham: Springer.

Cooper, R. G., & Edgett, S. J. (2012). Best practices in the idea-to-launch process and its governance. *Research-Technology Management, 55*(2), 43–54.

Frome, J., & Martin, P. (2019, August). Describing the game studies canon: A game citation analysis. In *Digital games research association conference 2019: Game, play and the emerging ludo mix.*

Gafni, R., Achituv, D. B., Eidelman, S., & Chatsky, T. (2018). The effects of gamification elements in e-learning platforms. *Online Journal of Applied Knowledge Management (OJAKM), 6*(2), 37–53.

Gauthier, A., Kato, P. M., Bul, K. C., Dunwell, I., Walker-Clarke, A., & Lameras, P. (2019). Board games for health: A systematic literature review and meta-analysis. *Games for Health Journal, 8*(2), 85–100.

Göbel, S., Vogt, S., & Konrad, R. (2018, October). Serious games information center. In *European conference on games based learning* (pp. 143–XVI). Academic Conferences International Limited.

Harpstead, E. (2017). *Projective replay analysis: A reflective approach for aligning educational games to their goals* (Doctoral dissertation, Carnegie Mellon University).

Harviainen, J. T. (2019). *Human–Computer Interaction and Simulation/Gaming*.

Isbister, K., & Mueller, F. F. (2015). Guidelines for the design of movement-based games and their relevance to HCI. *Human–Computer Interaction, 30*(3–4), 366–399.

Jackson, J., Iacovides, J., Duncan, M., Alders, M., Maben, J., & Anderson, J. (2020). Operationalizing resilient healthcare concepts through a serious video game for clinicians. *Applied Ergonomics, 87,* 103112.

Kayali, F., Luckner, N., Purgathofer, P., Spiel, K., & Fitzpatrick, G. (2018, August). Design considerations towards long-term engagement in games for health. In *Proceedings of the 13th International Conference on the Foundations of Digital Games* (pp. 1–8).

Kharrazi, H., Faiola, A., & Defazio, J. (2009, July). Healthcare game design: behavioral modeling of serious gaming design for children with chronic diseases. In *International conference on human–computer interaction* (pp. 335–344). Berlin, Heidelberg: Springer.

Kim, S. Y. S., Prestopnik, N., & Biocca, F. A. (2014). Body in the interactive game: How interface embodiment affects physical activity and health behavior change. *Computers in Human Behavior, 36,* 376–384.

Leung, R. (2014). Evaluating the benefits of collaboration in simulation games: The case of health care. *JMIR Serious Games, 2*(1), e1.

Mandigo, J., Lodewyk, K., & Tredway, J. (2019). Examining the impact of a teaching games for understanding approach on the development of physical literacy using the Passport for Life Assessment Tool. *Journal of Teaching in Physical Education, 38*(2), 136–145.

Mehta, N., & Pandit, A. (2018). Concurrence of big data analytics and healthcare: A systematic review. *International Journal of Medical Informatics, 114,* 57–65.

Mosca, I. (2017). What is it like to be a player? The qualia revolution in game studies. *Games and Culture, 12*(6), 585–604.

Mueller, F., Khot, R. A., Gerling, K., & Mandryk, R. (2016). Exertion games. *Foundations and Trends® in Human–Computer Interaction, 10*(1), 1–86.

Ohn, M. H., Ohn, K. M., Souza, U. D., Yusof, S., & Ariffin, Z. (2019, November). Effectiveness of innovative gamified learning among undergraduate medical students. *Journal of Physics: Conference Series, 1358*(1), 012060. IOP Publishing.

Olszewski, A. E., & Wolbrink, T. A. (2017). Serious gaming in medical education: A proposed structured framework for game development. *Simulation in Healthcare, 12*(4), 240–253.

Payne, H. E., Moxley, V. B., & MacDonald, E. (2015). Health behavior theory in physical activity game apps: A content analysis. *JMIR Serious Games, 3*(2), e4.

Ricciardi, F., & De Paolis, L. T. (2014). A comprehensive review of serious games in health professions. *International Journal of Computer Games Technology, 2014.*

Segura, E. M., & Isbister, K. (2015). Enabling co-located physical social play: A framework for design and evaluation. *Game user experience evaluation* (pp. 209–238). Cham: Springer.

Sobkow, B. (2017, May 9). Age rating for virtual reality games: Less restriction, more information. *More Information.*

Starks, K. (2014). Cognitive behavioral game design: A unified model for designing serious games. *Frontiers in Psychology, 5,* 28.

Tan, H., Wei, Y. C., Yun, H. W., Joan, K. E. H., Yee, H. W., & Juan, L. Y. (2020). Health en eTM: Developing a board game on value-based healthcare financing. *Simulation & Gaming, 51*(1), 87–105.

Verschueren, S., Buffel, C., & Vander Stichele, G. (2019). Developing theory-driven, evidence-based serious games for health: framework based on research community insights. *JMIR Serious Games, 7*(2), e11565.

Veselka, L., Wijesingha, R., Leatherdale, S. T., Turner, N. E., & Elton-Marshall, T. (2018). Factors associated with social casino gaming among adolescents across game types. *BMC Public Health, 18*(1), 1167.

Yap, K. Y. L., Yap, K. Z., & Yap, J. Y. G. (2015). A gamification framework for training of patient and medications management skills. *JILR, 1*(1), 47–58.

Zendle, D., & Scholten, O. J. (2020). The rise of social casino games: Changes in the size and composition of the Android social casino market from 2012–2020, availability to children, and predictions for future growth.

CHAPTER 5

Entertainment in Era of AI, Big Data & IoT

Giri Gandu Hallur, Sandeep Prabhu, and Avinash Aslekar

1 Introduction

The fundamental meaning of the verb "to entertain" points to providing pleasure, enjoyment, or diversion. "Entertain" has a supplementary meaning—an act of contemplation, deliberation, or musing (Mary Beth Oliver, 2014; Nadanyiova & Das, 2020). As a fact, the list of various activities that could be considered entertainment is non-ending and varies from person to person. The scope of possibilities is so expansive that it makes it implausible to present a succinct definition. Some researchers have attempted to provide a definition. Entertainment is a way of relaxing, recharging, rejuvenating, and unwinding (Haupert, 2006). It is a form of activity that engages an audience or gives pleasure and happiness. In

G. G. Hallur (✉) · S. Prabhu · A. Aslekar
Symbiosis Institute of Digital and Telecom Management, Symbiosis International (Deemed University), Pune, India
e-mail: girihallur@sidtm.edu.in

S. Prabhu
e-mail: sprabhu@sidtm.edu.in

A. Aslekar
e-mail: aslekar@sidtm.edu.in

S. Das and S. Gochhait (eds.), *Digital Entertainment*,
https://doi.org/10.1007/978-981-15-9724-4_5

the case of watching a play, opera, television show, or film; the audience may have a passive role, while it is active in the case of games (Haupert, 2006). Technology has its impact in almost all areas of human life and entertainment is no exception. The ongoing transition from live theater to the movies has advanced to home entertainment in the form of recorded/live music, radio, television, and has culminated into Internet-based entertainment (Haupert, 2006).

World Economic Forum in its 2016 white paper states that the growing global middle class will drive the uptake of innovative media offerings in the domains of education, premium multimedia content, and video-on-demand and the developing economies will be at the forefront of this adoption. The millennials are driving the demand for technology services that satisfy factors such as convenience, memorable experiences, and quick access to multimedia content (Singh, Mondal, Singh, Sahoo, & Das, 2020). Ever-increasing urbanization will fuel the increased demand for content services tailor-made to suit people's travel and busy professional lives (WEF, 2016). Advancements in handheld devices, cameras, rendering software, computing processors, smart TVs, earphones/headphones, advanced sound systems, availability of cheap sensors, cloud computing, and advanced wireless technologies have ushered the entertainment industry into a new era (Singh, Mondal, & Das, 2020). Enhanced device capability coupled with the use of technologies such as the Internet of Things (IoT), Artificial Intelligence, Social Media, Augmented reality (AR) and Virtual Reality and Big Data enable the entertainment industry to offer a range of seamless, personalized services (Das, 2021; Duy et al., 2020; Shukla, 2019; Van et al., 2020).

The World Economic Forum white paper highlights four key trends—Data analytics & real-time content management, adoption of smartphones, adoption of social media, and automation of content creation through the use of AI/ML (WEF, 2016). E &Y in its 2016 report on the Internet of Things states that IoT sensors will capture physical inputs and aggregate them as unprocessed data, which is then stored in digital format for analysis. It further adds that miniaturization has made it possible to embed sensors into smart devices thereby enhancing the role of IoT sensors from just data measurement to analytics by transmitting information via the internet (E & Y, 2016). So, as the various technologies converge today, IoT devices through sensors will capture large data that will be transferred via advanced internet connectivity to be processed and ingested by Big Data technologies and converted to actionable intelligence by AI/ML algorithms.

2 Literature Review

The literature review brings out three broad themes with reference to the application of AI, IoT, and Big Data in Entertainment:

1. Application of AI in Entertainment
2. Ingestion and Analysis of data collection IoT devices using Big Data
3. Interaction with the user through IoT devices.

2.1 Decision Making and Action Using AI

Review of Literature leads us to classify the application of AI in Entertainment into three subcategories:

1. Creation refers to the production of new media content by a creator.
2. Production consists of the way in which media content is edited and composed as delivered to the end consumer.
3. Consumption refers to the interaction an end consumer has with content.

A 2019 report by New European Media (Amato, 2019) does a comprehensive compilation of the AI tools and technologies used in the Media and Creative industries. As per the report, as is the case with Google's Deep Dream, deep learning is being used for producing art images as well as non-photorealistic images (Sharma & Das, 2020). The data-driven storytelling through advanced use of AI and ML has led to the development of 14 new digital storytelling tools (Das, 2020). Within the storytelling domain, immersive and interactive experiences are driven by sensors and data, enhancing the Quality of Experience (QoE) thus resulting in deeper engagement with the audience (Mondal, 2020). The use of knowledge representation, NLP, information gathering, and image forensics is being used to identify and dismantle fake news on social media and traditional media networks (de Lima, Feijó, & Furtado, 2020; Mondal & Sahoo, 2020) present an innovative approach to create individualized experiences based on the personalities of the users (Das &

Nayyar, 2020). The proposed method can finetune narratives according to the users' preferences and enhances the system's ability to deliver highly personalized narrative experiences (Das, Nayyar, & Singh, 2019). With artificial intelligence (AI) we can create virtual adaptive autonomous agents that can take various forms, depending on their surroundings (Das & Nayyar, 2019; Maes, 1995). The key areas of application of such a system can be videogames, movies, animation, etc. (Singh, Nayyar, Le, & Das, 2019). Chandler (2020) in his article stated that Warner Bros. has tied up with Cinelytic & developed an AI-enabled predictive system that can predict the probability of a film's success based on input parameters like actors acting in it, the budget of the movie and brands involved. Spyros (2018) in the case study presents cases of synergy AI, Big Data, and IoT. While IoT collects large amounts of data, AI is the proper tool to make sense of the Big data. AI can analyze, process, and bring out findings on which decisions can be made (Mohanty, Dash, Dash, & Das, 2019; Singh, Nayyar & Das, 2019). The AI-based analysis of Big Data enables us to identify patterns within the data so that we are able to make well-informed decisions (Gupta, Jena, Samantaray, & Das, 2019; Singh & Das, 2018). Google Home and Amazon's Alexa are examples of applications of AI, machine learning, and Big Data (Das, Mondal, Sahoo, Nayyar, & Musunuru, 2018; Jain, Jain, & Das, 2018). Yannakakis and Togelius (2018) in their book on the application of AI in Gaming, which is a popular application area. They go on to add that the ultimate goals of AI are social intelligence, human-like interactions, machine creativity, and general intelligence (Mondal, Das et al., 2017).

2.2 Ingestion and Analysis of Data Collection IoT Devices Using Big Data

Chang (2014) in their paper brings out that Big data technology and allied innovations related to data collection methodology, data mining algorithms, machine learning, and neural networks are changing all the businesses. May it be home entertainment, gaming, shopping, advertisements, mobile usage, or sports; every industry is finding a paradigm shift in their business function. Bravo and Marquez (Bravo-Marquez, 2014) state that Social media is an important communication means of the modern world. Consumers and the general public express their emotions such as happiness, anger, love, and hate through social media (Mondal, Mall et al., 2017). These comments are saved on social media servers as streaming data and are available for analysis for product brands (Das

& Mondal, 2016). Sentiment analysis is increasingly used by companies to understand the viewers' positive and negative polarised views for any movie as spectators express their preferences over social media. Researchers are proposing various methods to analyze and improve algorithmic performance. Lohrmann, Janacik, and Kao (2015) present in their publication the point that Event or streaming data require big data applications to ensure latency guarantee of the data optimizing on resource consumption. The big data model should be designed for efficiency of streaming service. In their 2016 paper, Smith and Telang (2016) state that the major technology disruption in entertainment is a big data technology and its streaming framework. Netflix pioneered the revolution by reworking its older website-based DVD ordering business model with big data model streaming movies, and soap operas directly on homes, mobiles, and other devices. Today, most of the video distribution industry runs on streaming big data business models. Yu and Wang (2015) state that sports fans express their emotions on social media in unstructured sentences. Fear, anger, anticipation, and joy expressed by these enthusiasts can be tracked by the sentiment analysis of social media statements. There is a predictive relationship between happening on the ground and sentiments streaming from social media. Rein and Memmert (2016) state that all sports analysts were dependent on observational data for deciding tactics for future games. Because of the availability of big data and analysis techniques in sports, advance planning for game tactics is possible. Grolinger, L'Heureux, Capretz, and Seewald (2016) conclude that the Internet of Things (IoT) devices are collecting huge time-series data and storing it in big data servers. Data collected through sensors installed at the event venues are analyzed using neural networks and deep learning algorithms. Li, Jiang, Yang, and Cuzzocrea (2015) conclude that organizations need a strategy not only to collect data but also to harness the data; to use data to study patterns, to achieve sustainable competitive advantage. Different algorithms based on machine learning and neural networks are useful to get important insight into the business activity. Big data strategy should comprise big data management systems, big data processing schemes, schemes for the application of big data streaming technology if required, big data privacy issues, and big data applications across different domains and departments of the organization. Hutchins (2016) states that technological advances in big data analysis in sports is offering numerous performance indicators to sports coaches, experts, and athletes. Even sport fans are delivered with many entertainment options in professional sports of this time. Safaei (2017) states that three V's of big data, namely, volume, meaning the large size of data availability, variety, meaning data type such as structured- semi-structured or unstructured, velocity, meaning data arrival rate- processing rate and data delivery rate.

2.3 *Interaction with the User Through IoT Devices*

Chen (2016) in 2016 paper summaries that the Internet of Things (IoT), big data, and artificial intelligence are advanced technologies that encompass the information and communication systems, but have found multiple applications across business, entertainment, and manufacturing. Cloud computing environment supports the process of big data and makes actionable decisions through the use of big data analyses and machine learning. Melber (2016) concludes her project on IoT by saying that IoT-enhanced entertainment satisfies a broad range of the key entertainment needs of consumers such as intuitiveness, convenience, and personalization. She adds that entertainment can be a key driver for smart home adoption through the use of IoT. Murphy and Popovici (Murphy & Popovici, 2014) in their paper suggest that users can get a new experience with toys through the use of the latest emerging wireless sensors (part of IoT), their interfaces managed through collaborative signal processing algorithms, and run on low-power networking hardware and protocols. They conclude by saying that the wireless sensor system allows the toys to respond to external stimuli, as part of the "internet of things," and paves the way for its application in interactive toys of the future. In their 2017 paper Kim, Park, and Choi (2017), describes the design and implementation process of IoT based systems for games environment that makes use of user's mobile devices and various IoT sensors. The paper presents the readers with the creation of games in the IoT environment by measuring and analyzing user interactions. Kosmides et al. (2018) discuss the possible use of the IoT environment to combine actions, decisions, and happenings in real-life with in-game progress through the application of advanced gaming technologies. The authors present the implementation of a gaming framework that uses Zigbee and Bluetooth Low Energy protocols as inputs to the system. They conclude by saying that the use of IoT will drive the deployment of new innovative IoT-based games and will create new market avenues for involved stakeholders. Jang and Bednarz (2018) through their publication based on the HoloSensor project demonstrate the analysis and application of visualization and visual analytics data received from IoT sensors embedded in Augmented Reality (AR) devices. They conclude that the combination of the immersive AR technology and IoT can find applications in the areas of personal health, smart homes, and entertainment. Dobrea and Dobrea (2017) present the concepts and the development process of an innovative wearable system that can monitor and analyze the position as well as the head movements of the wearer. They go on to add that this wearable has the potential to be applied in medical, entertainment, defence, automobiles, etc. When combined with AR technology this wearable system can be used in gaming. The entire system was designed to operate on low power and integrated with applications in the cloud. Iyer and Ozer (2016) explains the concept of Visual IoT that is a sub-domain of IoT that satisfies the need for sensing and processing of visual data. He also explains the modules involved (1) visual sensing nodes in the form of high-resolution cameras that capture the visual data perform initial processing locally (2) Gateways or platforms that save the visual data and also provide powerful computing capability, and (3) Cloud servers that index the data collected, provide analytics as well as storage. Khan, Bonné, and Shahid (2017) introduce the concept of a system that enhances interactivity through the use of smart objects and thus extends the interaction in case of advertisements. The IoT system connects several devices that are connected to a smart TV and enables cross-media interaction through the use of smartphones. The authors conclude that this concept of interactive advertisements can be used in animated movies, documentaries, etc. Zhao, Han, and Ma (2019) in their paper introduce a novel Virtual Reality based cartoon-style hybrid emotion personification model for live storytelling. The system attempts to enhance the audience's experience, by engaging with them through sharing of emotions thus making the experience memorable. During storytelling, the audience wears a VR headgear and explores the scene and watches the performance online thus enjoying an immersive VR experience.

The literature review brings out the following points:

1. AI has found application in the Entertainment industry in all three areas Creation, Production, and Consumption. In the areas of production, inputs from IoT sensors result in large amounts of data, which is then analyzed using Big Data analytics to arrive at actionable insights. In some cases, these data points are used as training

data for Machine Learning. Google Assistant, Alexa are examples of the convergence of AI, IoT, and Big Data.

2. AI with help from Big Data analyses the IoT data collected from sensors and provides highly accurate data of consumer behavior to analyze trends and patterns of purchasing.
3. Virtual Reality based emotion personification can enhance the audience's experience, by engaging with them through sharing of emotions thus creating an immersive VR experience.

3 Analysis of AI, Big DATA, & IoT Needs in the Entertainment Sector

The Entertainment industry is witnessing the convergence of device, network, and platform. Artificial intelligence is bringing a sea change in the way the entertainment industry is functioning. The penetration of AI technology in daily life is prevalent. AI Solutions for the entertainment industry are altering the way of functioning. We are soon going to witness a transformation of the entertainment industry in the nearest future as AI technology matures. The global AI software market is expected to head north increasing from $9.5 billion to $118.6 billion by the end of 2025 (Liu, 2020). As predicted by PwC, the entertainment and media industry in the USA is expected to touch $759 billion by 2021, growing at a compounded rate of 3.6% (PWC US edition, 2019). Just like other industries, big data is instrumental in technological disruption in the entertainment industry as well. It has effectively replaced older entertainment business models with new business models. Widespread adoption of OTT services, affordable yet feature-rich handheld devices, IoT- enabled devices and affordable broadband subscriptions support the consumer's demand for access to content anywhere- anytime. Given the wide range of choices for content, consumers have developed the paradox of choice. Moreover, intrusive and misdirected advertisements lead to fatigue among viewers, thereby rendering the investment on advertisements totally ineffective. So, the Entertainment industry's key challenges in today's era are how to create, deliver, and customize content across platforms. These challenges can be effectively overcome if real-time data of the consumers is collected and analyzed for bringing out patterns of the time and the context of media consumption.

4 Potential AI, Big DATA, & IoT Applications for the Entertainment Sector

With advancements in technology, AI, Big DATA, & IoT have found applications in motion films, animation movies, streaming services, sports, gaming & gambling, toys, and marketing and advertising. Just like other industries, big data is instrumental in technological disruption in the entertainment industry as well. It has effectively replaced older entertainment business models with new business models. By using sensors and actuators, IoT can provide users with highly responsive and immersive experiences that are far more engaging than the present-day entertainment; IoT can truly transform the future of entertainment.

4.1 Content Creation and Distribution

The use of real looking animated characters is also picking up where AI-generated characters play lead roles. In films like MUMMY use of CGI graphics is done, AVATAR had a mix of real and virtual characters. This movie used mixed reality. Some of the effects are based on a green screen. AI may find its applications here like Virtual Reality backed by AI. Disney is a pioneer in the collection of technical and creative properties like Marvel suite, Star War properties, and Pixar. Disney survived and grew with all digital characters by perfectly exploiting CGI. A humanoid AI robot Erica will play for the first time ever, an AI star in a feature film. Movies like Terminator: Dark Fate, Spider-Man: Far from home, I am a mother in 2019 have depicted AI.

The adoption of IoT has made it possible for firms to collect exponentially large quantities of data. Use of Big Data Analysis can help content companies to create content that can attract a wider audience than before. Over the years, Entertainment industry players have collected in-depth data by studying customer's behavior through their online interface. This analysis has enabled them to create original content for their target customer base. YouTube is a media sharing service website; while Netflix, Hulu, Prime Video, Apple TV+, Peacock, and Disney+ provide streaming services. Apple Music, Amazon Music, and Spotify specialize in music streaming; and Mixer and Twitch are platforms for video streamers. These media distribution companies have successfully built business models around streaming data. In 1997, "Netflix" meaning mix of "internet" and "flicks," introduced its first business model as a website-based movie rental business. Customers browsed the website and ordered DVDs, which were delivered through postal services. Once the renter views the movie, he would post it back. In 2010, Netflix moved on to a new business model consisting of streaming videos for consumers employing big data technology. Netflix's new business model was so formidable that the grand competitor "Blockbuster" filed for bankruptcy. Amazon, a major player in big data and entertainment industry started its journey as a bookseller; later moved on to build a big data network. Today, it leverages big data for its vast business portfolio from retail to entertainment services like Prime video and music. Netflix's acquisition of the production rights for the US of hugely popular British TV show "House of Cards" has been based on Big Data analytics. On a similar note, Amazon bought Kenneth Lonergan's highly acclaimed drama "Manchester by the Sea" as well as Whit Stillman's "Love and Friend."

4.2 Sports, Gaming, and Gambling

Sports have changed significantly from expert observational and opinionated data to big data analysis for organizing an event, communication, & execution of the event, as well as analyzing framewise data of actions of individual sportsmen on the field. Analysis of infield sports data generated by IoT sensors about players is helping coaches, athletes, and pundits in improving players' real-time field actions (Hutchins, 2016)—as well as team formations for tactical planning based on individual technical skills, the physiological performance of—one's own as well as opponent team. Sport analysts measure sentiments of viewers generated during the event using the flow of tweets during and after the event (Yu & Wang, 2015). Mobile sport tracking applications collect movement data to be used by city planners for designing cycling workout tracks. Event managers are increasingly using crowd sensing IoT and big data sensors to decide incentives for the event-goers. Big data analytics technology can enable real-time analysis of minute activities of users such as staying still, walking, running, and going up & going down. A typical soccer game analysis may include general statistical data to advanced video tracking analysis based on the coordinate system. The analysis may include passing behavior of the player, missed shot, space control, majority situations, player positions to acquiring the players.

The gaming industry is no child's play. Using wrist power of 2.5 billion gamers around the globe, the gaming industry is expected to reach $300 billion by 2025 (Koksal, 2019). Big data technology is the reason for the proliferation of online—real-time, multiplayer, multi-location games, and gaming platforms. Some of these game-changers in the gaming industry are Tencent, Sony, Microsoft, Apple, Activation Blizzard, Google, NetEase, Electronic Arts (EA), and Nintendo. Gamers create massive data trails as they drive through connected gaming platforms as they wore avatars, collect points, climb levels, and fight real-time opponents or race through fast roads. The massive data stream is managed through a data feed from big data servers and applications. Even the user data in terms of their interactions, duration and timings, opponents, and chat records are analyzed to customize user experience and engagement. Big data servers are designed to face varying loads and optimized performance. Gaming is emerging as a game-changer of the entertainment industry. Brain–computer interface headsets produce special effects through the use of brain signal processing algorithms based on a brain signal databank. Connected casinos enable automated gathering and analysis of reviews by gamers and real-time object detection during Entertainment events. These connected casinos gather data to analyze the customer behavior which can be used to anticipate player needs, make the game more appealing, or boost profits.

AI-enabled toys having ability to respond to voice and motion commands are getting common. These toys are equipped with machine learning software and can respond to voice using natural language processing. Some of them are Sphero BB-8, Codybot (a programmable robot) Vortex(a teaching robot) Meccano-Erector, hot wheels AI car, and many more.

4.3 User Experience Personalization

AI has an important role to play in user experience personalization. AI is being used to offer personalized content based on the data gathered from user activity and behavior on the website. Platforms like Netflix, Amazon are using AI to predict user choices and preferences. Contextual information is critical here, as the users' tastes or needs change depending on the time of day, their location, and the activity they are performing. Analysis of this contextual information should be followed by presentation of recommendations in a seamless user interface leading the user to his/her personalized content. The personalization can be achieved effectively if user data can be collected accurately and processed in real-time.

Uber through its tie-up with Spotify enables customers to connect their Spotify account to the Uber app. Customers can then access their playlist via the Uber app, and play music over the car's speakers while traveling. IoT offers solutions for data collection through use of various sensors embedded in mobile handsets, wearables, gaming consoles, etc. The Mob Museum, Las Vegas (DOPPELGANGSTERS, nd) teamed up with PwC Connected Solutions to create The Mob Museum mobile app that records the facial features of the user through a VR device and compares it with more than 800 images of heroes, villains, singers, comedians, entertainers as well as underworld leaders of the past. Users will be shown their match and the matches can easily be shared on social media using the mobile app. Another application of IoT is in the form of RFID-enabled wristbands for guests of a theme park that communicate with user's mobile phones. This application helps the theme park to identify the location of the guest, based on which, the IoT system can provide personalized entertainment based on a person's interests as he/she waits in line for a popular ride.

4.4 *Targeted Advertising*

Effectiveness of targeted advertising hinges on the availability of accurate user data, analysis of which gives advertisers a clear idea of preferences of the audience. The focus of Artificial intelligence programming will be on design and development of such advertisements. IoT sensors can provide the required volume, variety, and velocity of data, the Big Data and AI-based analysis of which can generate actionable insights for advertisers and broadcasters to create targetted marketing campaigns that are consonant with the target audience. This will help media companies to reach only those customers who have the intent to buy, leading to productive utilization of advertising expenses. Bentley has introduced the Bentley Network App (Bentley Network, nd) that leverages the connected car technology concept that opens up new business models involving partners such as Apple, Visa to offer its customers in the retail, maintenance, fuel, and infotainment services. In October 2019, Bentley has tied up with Viasat to provide drivers high-speed Wi-Fi from their cars through the app. Bentley offers its app partners data about its car users, their preferences, etc., helping them data required for targeted advertising. Google's local search ads through the use of IoT make it possible to target specific audiences with relevant ads within the vicinity of a certain location. The location data is fetched from the mobile phones as people make searches on Google. General Electric has introduced a software "Predix" (Predix, nd) that processes IoT data from smart machines in combination with traditional data to predict maintenance needs. Social Media companies such as Facebook, Amazon, Apple, and Google have introduced their own virtual assistant to tap into user data that can be processed to offer personalized and VR based immersive entertainment offerings. Media users often spend a lot of time searching for the content they are looking for and would prefer classified content. AI can accelerate the process of classification and make it more efficient. AI can do overall classification but classifying within favorites too. Moreover, recommendations based on search history make this more interesting bringing in the "Eureka" factor.

4.5 Enhancing User Experience Through VR and AR Through Immersive Content

Interactive video games along with music, movies, and television series have emerged as one of the most popular forms of entertainment. The traditional PC-based gaming is being replaced with the more interactive and engaging mobiles and consoles enabled gaming. This has driven the development of specialized sensor-enabled smart gaming devices which provide the user an immersive gaming experience. Adaptive storytelling via sensors in the face tracking module collects and analyzes the presenter's facial information in real-time and directly maps the detected emotion, eye state, and head position to the avatar. The presenter can also initialize the pre-configured multimodal effects via his keyboard and mouse. The audience wears a VR headgear and explores the scene and watches the performance online thus enjoying an immersive VR experience. Thus the combination of IoT and virtual or augmented reality can make content consumption completely immersive.

Successful application of AI, Big DATA, & IoT in the Entertainment industry set precedent for content companies to make key decisions regarding the type of content, the cast, target segment for the recommendation, and advertising channels through the use of these three technologies.

5 Drivers and Constraints for AI, Big DATA, & IoT in Entertainment

For the AI, Big DATA, & IoT to deliver to their potential, certain enablers would need to be in place and several constraints need to be overcome. These issues are technical, structural as well as behavioral. The regulatory issues concerning AI, Big DATA, & IoT need to be addressed, consumers need to develop trust in IoT-based systems and companies need to adopt the IoT-based data-driven approaches to decision making.

5.1 *Technology, Applications, and Infrastructure*

Building Artificial Intelligence mandates machine learning and deep learning techniques require a huge number of calculations to be made very quickly. Implementing AI requires a large amount of computing power. Recent developments in Quantum computing TFQ (TensorFlow Quantum) may help in this case however adoption of these technologies needs to be speeded up. Availability of AI-enabled big data applications in the entertainment industry is important for the ability of personalization, the autonomy of working, predictive ability, conversation ability, the ability of identification & anomalies, recognition ability, and objective orientation. NLP-based applications such as Amazon Alexa, Microsoft Cortana, and Apple Siri, are trained in two-way communication with customers using big data and AI algorithms. The technical robustness of an AI application is determined by the comprehensive tests it goes through. Building such test environments considering multiple possibilities is a challenging task. At times in haste to launch products some tests may be skipped causing unintended harm.

Conceptually big data technologies may sound simpler, but the technology implementation poses much bigger challenges. The Internet of Things (IoT) sensors produce massive streams of data, which at times needs to be collected, processed, and presented for consumption in real-time. Extensive knowledge and technological preparations are required to manage input and output load of data distribution. Unless you are a deep pocket large company, building and maintaining your own in-house big data infrastructure would be a challenge. Instead, outsourcing the service from some of the large outsourcing firms, is the best available solution. These firms not only provide data storage facilities but also provide support for data analytics and computing applications. The service provider takes care of all the hassles of flexible space requirements, data center management, power requirement, network infrastructure, data security issues, and technology up-gradation. These outsourcers are very flexible in their service offerings and easily scalable over the period of time. One can start small and gradually build-up the network. Some of the big international service providers in big data are Amazon Web Services S3, Microsoft Azure Data lake, Google Cloud Storage, Oracle Cloud, IBM Cloud, and Amazon Cloud (Marr, 2020).

For the large mass adoption of IoT in Entertainment, the cost of basic hardware must continue to drop. The low-cost, low-power sensors that are used in smartphones, AR/VR devices, smart TVs, smart home devices, and connected cars have seen a drop of up to 70% in the past five years, this trend should continue to drive the mass adoption of IoT. Innovation and deployment of high bandwidth and ubiquitous communication technologies such as Wi-Fi, 5G, and LoRa WAN, etc., will provide a boost to the adoption of IoT. Availability of affordable cloud-based or on-premise processing, open-source software tools, speech recognition, and datasets, innovation in analytical, visualization & content creation software and storage will provide the much needed backend support to the growth of IoT in Entertainment.

5.2 *Interoperability*

Adoption of interoperable standards and protocols will provide the much needed end-to-end interoperability of AI management standards, Big Data Analysis, immersive production, distribution, and consumption of media. Rob Claxton, the Chief Researcher at BT Research, explained that to create and manage automation in AI, the industry needs to build a framework to ensure the technology is used correctly, safely, and reliably—and that it is explainable and accountable.

Industry life cycle braves phases of startup, growth, shakeout, maturity, and decline. The big data industry has come a long way from startup to growth. In the growth phase every player, every technology pulls the industry in its own way leading to haphazard nonstandard development. Standardization in technologies of streaming, storage engine interface, querying, benchmarks, security & governance, metadata management, deployment, and other emerging technologies will bring uniformity in the industry. Though most of the big data applications are supported by the open-source Apache platforms; numerous Apache-based tools like Kafka,

Pulsar, Storm, Spark have their own interfaces. The Apache beam project is developed around the unification of these various tools. Other allied technologies also need significant standardization efforts for the seamless development of the big data industry.

The ability of IoT devices and systems to work together seamlessly is critical for realizing the potential of IoT applications. Adopting open standards is one way to accomplish interoperability. Several immersive media standards have been developed over the past few years but adequate effort to arrive at a consensus on new codecs, formats, protocols, APIs, etc. has not been taken.

5.3 Privacy and Confidentiality

AI has multiple legal dimensions when put to use for critical functions. Fixing responsibility for an accident caused by a self-driven car can be a big issue for insurance companies and transport authorities. Fundamental rights, prevalent regulations, values, and principles should be respected while developing and deploying AI applications. The definition of "ethical purpose" should be carefully done and implemented. The kind of data collected by IoT devices is large in volume and in most scenarios is very specific and hence, raises privacy and confidentiality concerns among users. Creators of IoT-enabled products & services offerings will have to create and convey to the users the value for the data collected and analyzed. They also need to be transparent about data collection and usage as well as ensure confidentiality. AI used for personalization requires the collection of user data over a period of time. Multiple AI applications may demand Personal, Medical, and Financial data too. The utilization of this data by the provider for other purposes is a matter of great concern. Although there are regulations like GDPR or PDP (in India) stringent implementation of such laws is still questionable. Privacy and confidentiality will serve as drivers or constraints depending on how well companies address end-user concerns regarding the management of their data.

5.4 Security

Adoption of AI, Big DATA, & IoT has exposed sensors and devices to the internet and the new types of security risks associated with it. So, IoT organizations while accumulating data from a large number of personal devices need to protect those data from unauthorized access. These companies also process and analyze these data in advanced IT systems located at remote locations, thus creating opportunities for possible data security breaches. So, management of security risks through adherence and compliance to industry best practices in data security as well as physical security of the IoT assets will serve as a driver for AI, Big DATA, & IoT adoption.

5.5 *Intellectual Property*

Content access being enabled across multiple devices has enriched consumer experience and led to increased consumption of content, but, as a downside it has opened up new opportunities for piracy. So, today content creators are in a constant struggle, to protect their IP as content makes its journey through multiple devices of authorized customers without lowering the quality of service. Protection of IP will need cooperation and collaboration of multiple stakeholders- regulatory and law enforcement agencies, content owners, Big DATA & IoT device/sensor manufacturers, AI application providers, internet service providers, content distribution networks, and consumers. This multi-stakeholder cooperation will go a long way in effectively protecting IP and drive further adoption of Big DATA & IoT in Entertainment.

5.6 *Public Policy*

Certain Big DATA & IoT applications such as connected or self-driving cars need approval from regulators. As technology evolves rapidly with applications across multiple sectors, regulators need to keep pace with enabling rules and regulations that enhance Big DATA & IoT adoption. Importantly, Government and regulators can play an enabling role in setting regulations for data collection, sharing, and use of data collected by connected devices and applications.

5.7 *Upskilling of Human Resources*

As the Entertainment industry enters the phase of largescale adoption of Big DATA & IoT, the human resources require a whole new set of multi-disciplinary skill sets and understanding & expertise in AI, Big DATA & IoT hardware and software. With the widespread acceptance of AR/VR and connected devices by customers, entertainment and content companies need to undergo a mindset change as they ideate, plan, and execute their content strategies to cater to immersive experiences. With the availability of enormous data, entertainment companies need to enhance skills in Big data analytics to unravel key insights about the customer. These insights will help to better leverage untapped incremental cross and up-selling opportunities.

6 Conclusion and Recommendations

Digital advancements will continue and AI will keep maturing. The actual challenge is to keep pace with development and handle the disruption it causes. AI will find its applications across industries and will disrupt each in different ways. Big data technology is extensively adopted by the entertainment sector. Big data architecture has streaming media capability which is exploited by the entertainment industry for streaming videos, music, and communication. Sports and event management industries have also successfully used big data technology for improving player and event performances. The gaming industry has built multiplayer, real-time games using big data technology. Big data also has important applications in Natural Language Processing (NLP) based on Artificial intelligence (AI) systems. Big data is administered and harnessed using the Hadoop ecosystem which works on the principle of distributed file system process, and then using frameworks such as MapReduce, HBase, Sqoop, Hive, Solr, Flume, and others. Currently the industry is dominated by large cloud-based service providers, outsourcing flexible, secure and scalable services to user organizations. The Internet of Things is fundamentally changing how content is created, distributed, and consumed. IoT enhances richness of the information collected and makes content consumption adaptive and immersive sharing. IoT has the potential to be much more than that. Management of security risks through adherence and compliance to industry best practices in data security as well as the physical security of the IoT assets will serve as a driver for IoT adoption. The industry needs to drive consensus on the adoption of interoperable standards and protocols that will provide the much needed end-to-end interoperability of AI management standards, Big Data Analysis, immersive production, distribution, and consumption of media. As industries enjoy the benefits of AI, Big DATA, & IoT they should be equally prepared for its undesirable negative consequences. Increasing AI, Big

DATA, & IoT deployments across disciplines will bring in legal, moral, and ethical issues. Appropriate regulatory measures by competent authorities after anticipating its impacts will help in striking the balance. Lastly, on the human resources front companies in the Entertainment industry need to upskill their employees in a whole new set of multi-disciplinary skill sets and understanding & expertise in IoT hardware and software.

References

Amato, G. B. (2019). *AI in the media and creative industries.*

Bentley Network. (n.d.). Retrieved from https://www.bentleymotors.com/en/world-of-bentley/ownership/technology/customer-mobile-applications.html.

Bravo-Marquez, F. M. (2014). Meta-level sentiment models for big social data analysis. *Knowledge-Based Systems, 69,* 86–99.

Chandler, S. (2020). *Hollywood is using artificial intelligence to Pick Its Next Blockbuster available.* Retrieved from Forbes.com: https://www.forbes.com/sites/simonchandler/2020/01/10/hollywood-is-using-artificial-intelligence-to-pick-its-next-blockbuster/#78bee3f775a5.

Chang, R. M. (2014). Understanding the paradigm shift to computational social science in the presence of big data. *Decision Support Systems, 63,* 67–80.

Chen, Y. (2016, June). Analyzing and Visual Programming Internet of Things and Autonomous Decentralized Systems. *Simulation Modelling Practice and Theory, 65,* 1–10.

Das, S. (2020). Innovations in Digital Banking Service Brand Equity and Millennial Consumerism. In *Digital Transformation and Innovative Services for Business and Learning* (pp. 62–79). IGI Global. https://doi.org/10.4018/978-1-7998-5175-2.ch004.

Das, S. (2021). *Search engine optimization and marketing: A recipe for success in digital marketing* (1st ed.). CRC Press. https://doi.org/10.1201/9780429298509.

Das, S., & Mondal, S. R. (2016). *Yours Digitally: Patanjali.* Available at SSRN 2737299.

Das, S., Mondal, S. R., Sahoo, K. K., Nayyar, A., & Musunuru, K. (2018). Study on Impact of socioeconomic make up of Facebook users on purchasing behavior. *Revista Espacios, 39,* 28–42. Retrieved from http://www.revistaespacios.com/a18v39n33/18393328.html.

Das, S., & Nayyar, A. (2019). Innovative ideas to manage urban traffic congestion in cognitive cities. In *Driving the development, management, and sustainability of cognitive cities* (pp. 139–162). IGI Global. https://doi.org/10.4018/978-1-5225-8085-0.ch006.

Das, S., & Nayyar, A. (2020). Effect of consumer green behavior perspective on green unwavering across various retail configurations. In *Green marketing as a positive driver toward business sustainability* (pp. 96–124). IGI Global. https://doi.org/10.4018/978-1-5225-9558-8.ch005.

Das, S., Nayyar, A., & Singh, I. (2019). An assessment of forerunners for customer loyalty in the selected financial sector by SEM approach toward their effect on business. *Data Technologies and Applications, 53*(4), 546–561. https://doi.org/10.1108/DTA-04-2019-0059.

de Lima, E. S., Feijó, B., & Furtado, A. L. (2020). Adaptive storytelling based on personality and preference modeling. *Entertainment Computing, 34,* 100342.

Dobrea, D. M., & Dobrea, M. C. (2017). *Concepts and developments of an wearable system—An IoT approach.* https://doi.org/10.1109/isscs.2017.8034922.

Doppel gangsters. (n.d.). Retrieved from https://themobmuseum.org: https://themobmuseum.org/press_releases/the-mob-museum-launches-experiential-mobile-app-designed-to-engage-visitors-find-their-doppelgangsters/.

Duy, N. T., Mondal, S. R., Van, N. T. T., Dzung, P. T., Minh, D. X. H., & Das, S. (2020). A study on the role of web 4.0 and 5.0 in the sustainable tourism ecosystem of Ho Chi Minh City, Vietnam. *Sustainability, 12*(17), 7140. https://doi.org/10.3390/su12177140.

E & Y. (2016). *Internet of Things Human-machine interactions that unlock possibilities Media & Entertainment.* Retrieved from ey.com: https://www.ey.com/Publication/vwLUAssets/ey-m-e-internet-of-things/$FILE/ey-m-e-internet-of-things.pdf

Grolinger, K., L'Heureux, A., Capretz, M. A., & Seewald, L. (2016). Energy forecasting for event venues: Big data and prediction accuracy. *Energy and Buildings, 112,* 222–233.

Gupta, D. K., Jena, D., Samantaray, A. K., & Das, S. (2019). HRD climate in selected public sector banks in India. *Revista ESPACIOS, 40*(11).

Haupert, M. J. (2006). *The Entertainment Industry* (G. Press, Ed.). Retrieved from Sciencedaily.com: https://www.sciencedaily.com/terms/entertainment.htm.

Hutchins, B. (2016). ales of the digital sublime: Tracing the relationship between big data and professional sport. *Convergence, 22*(5), 494–509.

Iyer, R., & Ozer, E. (2016, November–December). Visual IoT: Architectural challenges and opportunities; Toward a self-learning and energy-neutral IoT. *IEEE Micro,* 45–49. https://doi.org/10.1109/mm.2016.96.

Jain, S., Jain, V., & Das, S. (2018). Relationship analysis between emotional intelligence and service quality with special evidences from Indian banking sector. *Revista ESPACIOS, 39*(33).

Jang, J., & Bednarz, T. (2018). HoloSensor for smart home, health, entertainment (pp. 1–2). https://doi.org/10.1145/3213779.3213786.

Khan, V., Bonné, J., & Shahid, S. (2017). Interactive Advertisements in an IoT Era. *178*.

Kim, Y., Park, Y., & Choi, J. (2017). A study on the adoption of IoT smart home service: Using Value-based Adoption Model. *Total Quality Management & Business Excellence, 28*(9–10), 1149–1165.

Koksal, I. (2019). *Video Gaming Industry & Its Revenue Shift*. Retrieved from Forbes.com: https://www.forbes.com/sites/ilkerkoksal/2019/11/08/video-gaming-industry–its-revenue-shift/#48161f30663e.

Kosmides, P., Demestichas, K., Adamopoulou, E., Koutsouris, N., Oikonomidis, Y., & De Luca, V. (2018). *Inlife: Combining real life with serious games using IoT*. https://doi.org/10.1109/cig.2018.8490434.

Li, K. C., Jiang, H., Yang, L. T., & Cuzzocrea, A. (2015). *Big data: Algorithms, analytics, and applications*. CRC Press.

Liu, S. (2020). *Artificial intelligence software market growth forecast worldwide*. Retrieved from statista.com: https://www.statista.com/statistics/607960/worldwide-artificial-intelligence-market-growth/.

Lohrmann, B., Janacik, P., & Kao, O. (2015, June). Elastic stream processing with latency guarantees. In *2015 IEEE 35th International Conference on Distributed Computing Systems* (pp. 399–410). IEEE.

Maes, P. (1995). Artificial life meets entertainment: lifelike autonomous agents. *Communications of the ACM*, 108–114.

Marr, B. (2020). *What are the best big data cloud storage providers? Here are the top 6 available at*. Retrieved from Forbes.com: https://www.forbes.com/sites/bernardmarr/2020/05/11/what-are-the-best-big-data-cloud-storage-providers-here-are-the-top-6/#6d254f265cd2.

Mary Beth Oliver, A. A. (2014, June). An Introduction to the Special Issue: Expanding the Boundaries of Entertainment Research. *Journal of Communication*, 361–368.

Melber, L. F. (2016). *IoT-enhanced Entertainment: A new use case for 5G Fixed Wireless Access)*. Retrieved from semanticscholar.org: https://www.semanticscholar.org/paper/IoT-enhanced-Entertainment%3A-A-new-use-case-for-5G-%3A-Melber/907f95f65fe525cf9d90455cf6d836766c79104e.

Mohanty, P.C., Dash, M., Dash, M., & Das, S. (2019). A study on factors influencing training effectiveness. *Revista Espacios, 40,* 7–15. Retrieved from http://www.revistaespacios.com/a19v40n02/19400207.html.

Mondal, S. R. (2020). A systematic study for digital innovation in management education: An integrated approach towards problem-based learning in Vietnam. In *Digital innovations for customer engagement, management, and organizational improvement* (pp. 104–120). IGI Global. https://doi.org/10.4018/978-1-7998-5171-4.ch006.

Mondal, S., Das, S., Musunuru, K., & Dash, M. (2017). Study on the factors affecting customer purchase activity in retail stores by confirmatory factor analysis. *Revista Espacios*, *38*(61).

Mondal, S., Mall, M., Mishra, U. S., & Sahoo, K. (2017). Investigating the factors affecting customer purchase activity in retail stores. *Revista Espacios*, *38*(57).

Mondal, S., & Sahoo, K. K. (2020). A study of green building prospects on sustainable management decision making. In *Green building management and smart automation* (pp. 220–234). IGI Global. https://doi.org/10.4018/978-1-5225-9754-4.ch011.

Murphy, F., & Popovici, E. (2014). *Design and implementation of a heterogeneous, power efficient wireless sensor network for smart toys*.

Nadanyiova, M., & Das, S. (2020). Millennials as a target segment of socially responsible communication within the business strategy. *Littera Scripta*, *13*(1), 119–134. https://doi.org/10.36708/Littera_Scripta2020/1/8.

Predix. (n.d.). Retrieved from ge.com: https://www.ge.com/digital/iiot-platform.

PWC US edition. (2019). *Global Entertainment & Media Outlook 2019 2023*. Retrieved from pwc.com: https://www.pwc.com/us/outlook.

Rein, R., & Memmert, D. (2016). Big data and tactical analysis in elite soccer: Future challenges and opportunities for sports science. *SpringerPlus*, *5*(1), 1–13.

Safaei, A. A. (2017). Real-time processing of streaming big data. *Real-Time System, 53*(1), 1–44.

Sharma, E., & Das, S. (2020). Measuring impact of Indian ports on environment and effectiveness of remedial measures towards environmental pollution. *International Journal Environment and Waste Management, 25*(3), 356–380. https://doi.org/10.1504/IJEWM.2019.10021787.

Shukla, D. (2019). *Technology Trends Transforming Media and Entertainment Industry*. Retrieved from electronicsforu.com: https://www.electronicsforu.com/technology-trends/technology-trends-transforming-media-entertainment-industry.

Singh, S., & Das, S. (2018). Impact of post-merger and acquisition activities on the financial performance of banks: A study of Indian private sector and public sector banks. *Revista Espacios Magazine, 39*(26), 25.

Singh, L. B., Mondal, S. R., & Das, S. (2020). Human resource practices & their observed significance for Indian SMEs. *Revista ESPACIOS*, *41*(07). Retrieved from http://www.revistaespacios.com/a20v41n07/20410715.html.

Singh, S., Mondal, S., Singh, L. B., Sahoo, K. K., & Das, S. (2020b). An empirical evidence study of consumer perception and socioeconomic profiles for digital stores in Vietnam. *Sustainability*, *12*(5), 1716.

Singh, I., Nayyar, A., & Das, S. (2019). A study of antecedents of customer loyalty in banking & insurance sector and their impact on business performance. *Revista ESPACIOS*, *40*(6).

Singh, I., Nayyar, A., Le, D.H., & Das, S. (2019). A conceptual analysis of internet banking users' perceptions. An Indian perceptive. *Revista ESPACIOS*, *40*(14).

Smith, M. D., & Telang, R. (2016). *Streaming, sharing, stealing: big data and the future of entertainment*. MIT Press.

Spyros. (2018). Synergy of IoT and AI in Modern Society: The Robotics and Automation Case. *Robot Autom Eng J*, *3*(5).

Van, N. T. T., Vrana, V., Duy, N. T., Minh, D. X. H., Dzung, P. T., Mondal, S. R., & Das, S. (2020). The role of human–machine interactive devices for post-COVID-19 innovative sustainable tourism in Ho Chi Minh City, Vietnam. *Sustainability*, *12*(22), 9523. https://doi.org/10.3390/su12229523.

WEF. (2016). *Digital Transformation of Industries: Media, Entertainment and Information*. Retrieved from reports.weforum.org: http://reports.weforum.org/digital-transformation/wp-content/blogs.dir/94/mp/files/pages/files/wef-dti-mediawhitepaper-final-january-2016.pdf.

Yannakakis, G. N., & Togelius, J. (2018). *Artificial intelligence and games* (Vol. 2). New York: Springer.

Yu, Y., & Wang, X. (2015). World Cup 2014 in the Twitter World: A big data analysis of sentiments in US sports fans' tweets. *Computers in Human Behavior*, *48*, 392–400.

Zhao, Z., Han, F., & Ma, X. (2019). *Live emoji: A live storytelling VR system with programmable cartoon-style emotion embodiment* (pp. 251–2511). https://doi.org/10.1109/aivr46125.2019.00057.

CHAPTER 6

Digital Application-Based Viewing: Impact of Instructions & Feedback on Viewer Satisfaction & Learning Effectiveness

Sangeeta Pandit

1 Introduction

Digital can be described as an electronic technology. It is an improvement over analog technology that carried waves of given frequency (Nadanyiova & Das, 2020; Singh, Mondal, Singh, Sahoo, & Das, 2020). Satellite, fiber optic transmissions, and finally a modem are now the primary methods of digital transmission. Digital Applications can be said to be utilities or targeted uses that are made available through internet connection (Singh, Mondal, & Das, 2020). End users, of digital applications, are those that view content with the help of the internet (Sharma & Das, 2020). We view content on our television, computers be they laptops or desktops, smartphones, kindle, and tablet. Content can be interactive like games and gambling (Das, 2020).

S. Pandit (✉)
SIMSREE, Mumbai, India
e-mail: sangeeta@pdkunte.com

S. Das and S. Gochhait (eds.), *Digital Entertainment*,
https://doi.org/10.1007/978-981-15-9724-4_6

OTT (Over the Top) are services received through internet. Netflix, Amazon Prime, and Hotstar are some such examples (Mondal & Sahoo, 2020). The viewer has vast choice of content that can be watched anywhere, anytime without advertisements but internet connection is needed (Mondal, 2020). Television came to India, in early seventies. It gradually evolved into many channels and even big screen stars started realizing the power and popularity of television mode (Das & Nayyar, 2020). Some serials became very popular with a large number of people being glued to their television set, so much so, that in that period traffic on the roads reduced considerably (Das & Mondal, 2016). News channels besides giving news also started becoming influencers having debates and taking positions on various important National and International issues (Das, Nayyar, & Singh, 2019). In India, even in the eighties, it was difficult to get a telephone connection. There was lot of red tapism and a long wait. But soon, their need reduced with introduction of mobile phones in late nineties (Das & Nayyar, 2019). Launch of cordless telephones charmed consumers very briefly (Singh, Nayyar, Le, & Das, 2019). The ubiquitous Nokia mobile phone saw a customer base including the length and breadth of India and across all economic classes from the poor to the super rich. For many, it was a support to their business (Mohanty, Dash, Dash, & Das, 2019). Somewhere around 2013, Samsung with their dual sim card feature started shrinking the Nokia market. Camera became an important feature of mobile phones (Singh, Nayyar, & Das, 2019). 2G enabled messages services and multimedia messages which became a marketing tool (Singh & Das, 2018). 3G increased bandwidth and speed. People became comfortable using email feature on phones (Gupta, Jena, Samantaray, & Das, 2019).

4G exponentially increased bandwidth and with that speed and access across countries (Jain, Jain, & Das, 2018). With fierce competition amongst smartphone players, costs started falling and large number of Indians started using smartphones. In fact, most people using smartphone in India have never used computers (Das, Mondal, Sahoo, Nayyar, & Musunuru, 2018). Though internet came to India in the nineties, it became popular with the masses only when smartphones became relatively cheap (Mondal, Das et al., 2017). The smartphones with their various additional features like Bluetooth and being able to access videos, download apps, and so on are now used aggressively (Mondal, Mall et al., 2017). Table 1 presents a snapshot of digital numbers in India as on May, 2018.

Table 1 Digital numbers in India on May 2018

Digital identity (Aadhar)—1.2 billion
Internet users—450 million
Mobile phone users—1.18 billion
Smartphone users—298 million devices and 468 million connections
Social media users—250 million
Books—0.7 million books in 70 languages in National Digital Library
Ration cards—100% digitized
Govt eMarketplace (GEM 3.0)—263,731 products by 105,889 sellers listed, with 22,598 buyer organizations

Source www.digitalindia.gov.in (Ministry of Electronics & Information Technology, Government of India)

Digital Industry is a sunrise industry, there are various sectors in it that overlap and are interconnected: digital media, video on demand, games, gambling, social media, streaming services, e-tailing & e-commerce, tools, apps, and so on. List is long and industry is still in the nascent stage. Data is the new wealth, digital tools and apps, the new assets, or weapons to survive and excel. The benefits of digital applications are huge but the risks too are high. New kind of criminals have emerged. Intelligent, difficult to track, some with motive of pecuniary gain, some paid to influence gullible people, and some just psychopaths. One must be mindful of the risks and all stakeholders of the digital system sensitized to protect themselves.

The Digital Applications Eco-System comprises of Content Generators, Content Creators, and Content Developers. There are those that enable content and consume content and they are dependent on each other. Content creators are many like Governments, Business Bodies, NGOs, Institutes, and Individuals. They are aided and supported by Website Designers, Application Developers, Content Management System creators, and the like. Machine Translators and Tool Developers play a vital role. Microsoft tools, Jira, Trello, and many more are widely used. Consumers have products like Chat, Entertainment, News, Advertisements, Marketing sites, Payment ways, Social Media Platforms, and so on. Obtaining Government permissions, payment of taxes, banking procedures, and many services have become transparent and efficient due to digital technology. Digital Applications is due to Internet penetration. Internet access is reaching the villages and remote areas of India. Figure 1

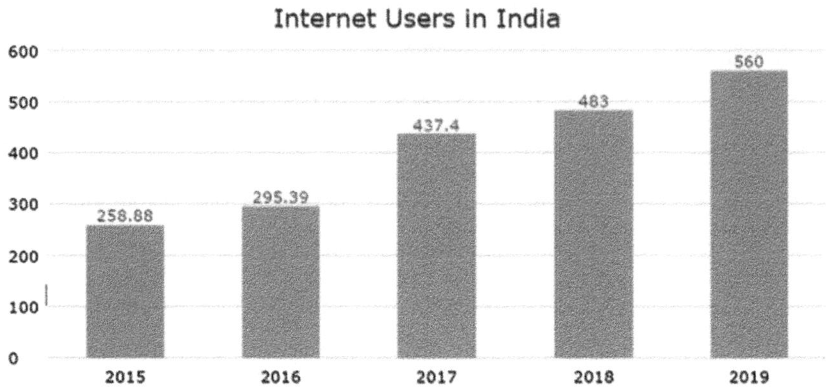

Fig. 1 Internet users in India (Numbers in billions) (*Source* https://sannams4.com/digital-and-social-media-landscape-in-india/)

gives a perspective of how digitization is inevitable and must be accepted.

1.1 Purpose of Chapter

This research study aims to understand the impact of digital application-based viewing. Its impact is due to its format. Its effect on viewers is in regard with satisfaction and desired results. It endeavors to find how this could be one of the solutions to help in nation building.

2 Literature Review

Nitecki (2019) in his post has discussed how an organization can lose market share though product is good and marketing team expertise is excellent. Technologies have changed landscapes, and mobile application is the new way to growth. Digital transformation is needed to survive. It improves customer experience and accessibility. Organizations need to change their culture and make a long-term commitment to digital transformation. Crawford (2017) has summarized the Gartner report on Digital Applications by explaining that companies are changing their purchasing patterns due to emerging digital strategies. Demand for innovations that are technology based that give solutions to monetizing the value delivered is on the increase. Application providers too are constantly adjusting to changing business needs. Software providers are competing to help businesses evolve toward digital business.

Luna-Nevarez and McGovern (2018) have, in their research paper on digital pedagogy on higher education, concluded that it is important for educators to understand and adopt to student's digital lifestyles. Many Institutes are providing online content to students. However, a faculty must customize content by curating a digital magazine just for his or her subject. The magazine could contain snippets of news and stories relevant to concept and topic to be delivered in class, maybe from the sites of e newspapers, articles from Journals, test book chapter summary, and so on. Student's engagement and learning outcome was better than just reading online pdf content. Faculties must accept the benefits of digital learning. Digital curated magazines are an excellent tool to use to enhance student learning and be a better educator. It is imperative that faculties learn to use mobile apps to keep students engaged (Das, 2021; Duy et al., 2020).

Agarwal and Karahanna (2000) in their article have discussed the aspects of cognitive absorption and technology use. Creativity and fun-loving traits of a person are necessary for cognitive absorption. Digital applications must be perceived to be useful and easy to use. The experience of digital application must be such that one is totally involved, is enjoying, not attention wandering, and curiosity is invoked (Venkatesh, Thong, & Xu, 2012). Syrjamaki, lsokoski, Surakka, Pasanen, and Hietanen (2020) in their article on eye contact reported findings of their experiment on psychophysiological impact of eye contact: if it differed and to what degree when it was virtual and when it was face to face. Direct gaze had bigger impact than averted gaze. The effect on heart rate deceleration was about the same both in virtual and physical situations. However, physiological effects like skin conductance responses were more in physical direct look rather than virtual direct look.

Andaç, Akbiyuk, and Karkar (2016) have discussed in their article the importance of creative content. Visuals and auditory contents can invoke customer bonding. Content must be effective and reliable. Digital content marketing tool can influence a customer greatly if it communicates well understanding the customer. It is more a way to be in touch with current and potential customers. Alkhafaji (2016) has studied the impact of mobile apps on customer satisfaction. Businesses need to promote their product, and mobile apps is one of the best modes. Mobile apps can be made interactive. Customer relations improve and CRM factor is vital for a business success. The study showed how mobile apps are beneficial both to sender and receiver. However, as they are more effective on smartphones, a version could be developed for mobile phones that are not smartphones.

Lundberg and Westerman (2020) have in their article referred to effective Chief Learning Officers as Transformers. One of the issues they have discussed is how transformers need to create digital awareness and develop digital aptitude of employees. One need not be very technically skilled, but people must be comfortable in digital space. Data-driven thinking is the way forward. Employees as per their profile should receive relevant processed data and use it to augment bottom line and their performance ratings. Data is of great value and it can challenge accepted way of thinking (Van et al., 2020).

Govindrajan and Srivastava (2020) have in their article tackled the issue of need of 4-year-long residential program, level of infrastructure needed by Institutes to deliver online education, and the type of training and changing of mindsets needed of both faculties and students. They have reasoned that technology platforms can enable students to learn at their own pace at low cost. Certain parts of curriculum can be commoditized and delivered with same effectiveness whether online or face to face. Group interactions, mentoring, fostering team spirit, social skills, and the like could be the focus of face-to-face classes. This hybrid model will free time for meaningful activities that need physical presence. Level of infrastructure will differ from Institute to Institute. Those having better infrastructure and financial muscle to invest in the same will have an edge. Also, students having better appliances will be at an advantage. The COVID 19 pandemic has initiated a global experiment in online education. Tectonic shifts in society have always churned out solutions. This pandemic too may result in innovative solutions to have effective communication through digital mode.

Rajat Basu in his article in ET dated 10th June 2010 discussed the PWC Global Entertainment and Media Report. India is poised for tremendous growth in the OTT sector: Video OTT to reach a staggering \$1.7 billion by 2023; and subscription video on Demand to fetch revenue of \$1.5 billion. Traditional TV will stay, paid TV penetration is good, and it will not decrease. TV advertisement revenues will increase. However, there will be a boom in digital advertisements, from a \$1.2 billion industry in 2018 it will grow to \$2.6 billion by 2023. Insead Business School website discusses development of effective strategies for the digital age. Cloud, AI, and the like that are all interconnected open both opportunities and risks. Products, services, and business models will continually use improved digital applications. Strategy must be agile to adapt to change caused by disruption. Environment today is not predictable, time frame for disruptions cannot be predicted, and business must develop effective strategies to absorb benefits of digitization. Tables 2 and 3 explain the literature and variables for consideration in detail.

Table 2 is prepared after perusing the literature. The gaps are noted, and variables identified to help prepare the questionnaire for collection of primary data.

Digital Transformation must be accepted by organizations irrespective of size and nature. Products and services to be marketed need

Table 2 Prominent literature considered

Author	*Year*	*Published in*	*Area of research*	*Gap*
Nitecki	2019	Blog-Netguru SA	Mobile Applications-Marketing Tool	Exploring other digital tools
Crawford	2017	Blog-Gartner Network	Digital Innovations are prodded by business demands	Opportunities shown by software & technology innovators
Nevarez, McGovern	2018	Journal of Educational Technology Systems	Curated Digital Magazine	Skillsets needed and Resources available
Agarwal, Karahanna	2000	MIS Quarterly	Cognitive Absorption	Time period of interaction
Syrjamaki, Isokoski et al.	2020	Journal of Computers in Human Behavior	Eye Contact	Training of speaker & listener
Andac, Akbiyik, Karkar	2016	Journal of Business Research	Effective Content	Requirements for Content to be effective
Alkhafaji	2016	Archives of Business Research	Mobile apps in mobile phones that are not smartphones	Making mobiles that are not smartphones also a marketing tool
Lundberg, Westerman	2020	HBR	Digital Awareness & Attitude	Way to create the digital awareness
Govindrajan, Srivastava	2020	HBR	Infrastructure and appliances matter	How to overcome the problem

Table 3 Table of variables

Independent variable	*Dependent variable*	*Moderating variable*
Business Survival and Growth	Market Share	Mobile App
Digital Innovations	Business Needs	Creativity of innovators
Student Learning	Digital Magazine	Faculty Expertise
Cognitive Absorption	Digital technology	Fun and easy to use
Impact on audience	Eye Contact	Direct Gaze
Customer Bonding	Communication	Mobile App
High-Performing Employees	Data-driven thinking	Digital Training
On-Line Education	Digital Platforms	Infrastructure and appliances

digital support. Cost, efficiency, effectiveness, and accessibility exponentially increase, and those who cling to traditional methods will lose market share. Market leaders will be replaced by organization's who use digital tools. Digitization not only disseminates information faster and to a wider audience, it also influences people, builds relationships and the various social media platforms subtly capture consumer's fancy. Technology has pervaded every aspect of our life. Encyclopedias, dictionaries, and atlases are replaced by Google search. Extent of information available on the internet on every conceivable topic is mind boggling. Businesses need to leverage technology for growth. Every organization is unique. Each has its own DNA of culture, type of leadership, level of skillsets of employees, work environment, and so on. Digital needs too will differ from organization to organization. Management must identify their digital needs and get it met with the in-house technology team or with the technology innovators in the market. The COVID 19 pandemic has made online education a necessity. Sessions delivered online to students need to have certain features to keep the students engaged and most importantly result in learning. Instead of just PDF files, curated digital magazines given as pre-reading material may have a better impact. Cognitive is the process of thinking and reasoning. Cognitive ability increases with visuals over words. Online material must invoke thinking, creativity, and understanding. It must be attractively designed and user-friendly. In face-to-face physical interactions, voice modulation, body language eye contact, and touch are some aspects that can greatly influence the other person or persons. In virtual mode, the same impact can be generated by eye-to-eye contact and by adoption of other digital etiquettes. To digitally transform organizations, mindsets of people must change. Cloud, AI, and social media platforms are powerful tools that need to be used beneficially in a win-win way. Educational Institutes that do not have funds available for updating could lose the race in the digital era. They need to spend on library, training of faculties and investing in technology platforms. Students from lower strata of society if not possessing smartphones or computers and not having access to internet would suffer.

3 Research Methodology

Questionnaire and Interview method was adopted. Based on the Table of Variables constructed, questionnaire was designed. It was sent to 100 people based in Mumbai and Pune. A Random Sampling method was followed. Age group was from 25 to 55. 60% were males and 40% were females. Income levels ranged from Rs. 180000 per annum to Rs. 2500000 per annum. Interviews were taken of 25 people of varied professions being performers, content creation teams, technology people creating applications, and entertainment receivers. The study was conducted in the month of May 2020.

4 Findings

Following is the summary of their responses.

Majority of the recipients had smartphones. A few possessed mobile phones with only phone and messaging feature, but as a second device, to be used in emergency if the smartphone breaks down. For 40% of the respondents' smartphone purchase was a major expenditure and they had to save money to buy it over an average period of 6 months. 20% of the respondents had taken loan or purchased on EMI. For artisans and laborers, it was a necessity. It saved time and travel, as photos of tools, designs could be sent to customers for approvals or shops for enquiring availability. For shopkeepers, it helped in increasing sales, as they could cater to physical customers and at the same time receive orders from customers on message and WhatsApp. Apps are very popular. In Entertainment, popular are Gaana and Saavn; in Shopping—Flipkart and

BigBasket; in Food—Zomato and Swiggy; in Transportation—Ola and Uber; and for Payment options—Paytm. Most have Facebook and Instagram too. People use apps extensively to play games, watch videos, listen to music, watch live matches, and to be updated on current happenings. Facebook and Instagram are popular with all people from all economic backgrounds. People are ready and eager to help each other to explore usage of smartphone features. Apps are very smooth and easy to use and reduce streaming time. People are earning money by uploading videos on YouTube of them singing, dancing, and sharing tips on gardening, farming, cooking, and so on; if they become popular, watched by many, advertisements appear and they receive a share of the advertisement revenue from the digital platform. People from humble backgrounds, with no technical education, have grasped the nuances of digital applications. They share their expertise on topics like farming and cooking and many have huge fan following and become influencers too. All the smartphone users use the phone every day to listen to music or watch videos. In this COVID 19 situation, people have attended webinars and conducted them using platforms like Bluejeans, WebEx, ZOOM, and Google Meet. Schools have conducted online sessions with parental supervision for younger children. Institutes delivering Graduation courses and Higher Education Institutes have conducted online sessions. People from all backgrounds, age groups, and profiles whether students, professionals, homemakers, entrepreneurs, and so on have taken benefit of online content to upgrade themselves. Most went for free content and just a few have paid to attend online sessions. Statistically, there was no relationship between income levels and digital application users. There also was no relationship between education levels and digital application users.

5 Recommendations

India with its humongous population can leverage digital applications for social welfare, education, entertainment, and other purposes. The extent and level of usage could be exponential. Many big players like Amazon and Netflix have realized it and invested in India in a big way. Many start-ups too have mushroomed. With internet speed and access increasing, digital applications have a bright future. Reliance Jio has revolutionized data usage and with attractive schemes increased the numbers of new mobile users and captured the market from their competitors.

Content in regional languages has also attributed to much increase. The KPMG Google report of Indian language Internet users reported that in India, 68% of the Internet users consider regional language digital content to be more reliable. Maximum regional language internet use is of Tamil language followed by Kannada, Bengali, Marathi, Telegu, Gujarati, and Marathi. It can service the agricultural sector in a very meaningful way. AI, Cloud, block chain, social media, mobile apps, and the like are here to stay. They are going to see increase in features and number of users. India will soon be the most populated country in the world with a median age group of people in the thirties. Exposure to free content has created aspirations in people. If they are not adequately met, it could lead to social unrest. C. K. Prahalad has famously said that India is heading toward demographic disaster or dividends. Digital Applications must be perceived as a social impact tool and be used to spread education, awareness against social evils, open opportunities for businesses of SMEs & MSMEs to prosper and to reduce the gap between the rich and poor. Digital Applications may disrupt established sectors like Bollywood, Academic Institutes with sprawling campuses, Advertising channels, and others and be replaced by customized, quality, cheap, accessible, and user-specific needs content. Government of India embraced digitization and a lot of work is done online. It has led to decrease in corruption and increase in transparency. Banking functions and stock market operations use digital tools aggressively.

The most important findings of this research are that digital applications if used prudently by all stakeholders can educate, entertain, give employment, and be used as a tool for social welfare. If misused, it can lead to increase in cyber-crime, people being influenced negatively, using time unproductively, and becoming unfit for employment. RBI for banks, SEBI for stock markets, AMFI for mutual funds do play a role to ensure ethical practices. So also, an apex body is needed of ethical people to ensure that the benefits of digital applications reach the last mile and risks associated with them are minimized. Laws to regulate this dynamic virtual industry would be difficult; it could backfire stifling creativity and growth. What we need is good leadership, people committed to ethics and value system to self-regulate this sector. Science fiction is becoming a reality. We need heroes to eradicate and defeat the dangerous elements and criminals, petty and big. We must stop cyber-crimes like identity theft and others and most importantly maliciously influencing impressionable young minds. Digital tools like Aadhar card and Jandhan Yojana with the

Table 4 Digital applications generic features

For user	*For supplier*	*For Software developers*
Easy to Use	Business Growth	To understand needs of business
Fun element	Increased Profits	To understand needs of ultimate user
Evokes positivity	Customer Loyalty	To protect privacy
Educative	Ethical Factual Content	Agility to adapt and re adapt to changing needs
Initiates thinking	Transparency of Intent of Content	Sensitive to culture and beliefs of people
Develops creativity	Value addition to customer	Cyber security

support of other sophisticated tools could make India truly reap Demographic dividends. In the words of Mahatma Gandhi "Poverty is the worst form of Violence." Digital Applications must be used to bridge the gap between the rich and poor; the digital application eco-system must be inclusive bringing into its fold people from all economic backgrounds. Table 4 presents a generic check list for digital applications.

6 Conclusion

The COVID 19 pandemic has made many realize the power of digital mode. Online meetings reduced expenditures on travel, food, rent, and electricity. Employees could utilize time saved on travel for their hobbies and passions. Students can study at their own pace. Stress and peer pressure are reduced. Those not digitally inclined are compelled to accept it. Faculties started understanding student needs and accordingly curating session deliveries. Priorities have changed. Yoga, meditation, and exercise videos are on demand. People are spending time on exercising. Purchases of food items cooked and uncooked are increasing and of apparel are decreasing. Skill-upgrading courses whether it be investment strategy, computer languages or gardening, hair cutting, and the like are becoming very popular. People not highly educated of rural and urban India are comfortable with digital mode. The downside of use of digital applications is a vague feeling of being disconnected with reality and the possibility of slipping into depression. Getting addicted to games and watching videos at the cost of not completing important money-earning tasks. Reading information that is false and misleading, that hampers positive way of thinking. Risk of hacking is tremendous. Cyber crime is on the rise. Surprisingly 60% of the respondents had lost their smartphones, either stolen or misplaced and were traumatized by the loss. 40% had suffered due to excessive use with issues like carpal tunnel syndrome, finger cramping, back ache, or strained elbow due to wrong posture, blurred vision, or eyes watering. 30% of the respondents were very uneasy and nervous if their smartphone was not near them.

References

Agarwal, R., & Karahanna, E. (2000). Time flies when you're having fun: Cognitive absorption and beliefs about information technology usage. *MIS quarterly*, 665–694.

Alkhafaji, S. M. (2016). User satisfaction on mobile apps: An analytical study on Omani business environment. *Archives of Business Research*, *4*(1).

Andaç, A., Akbiyuk, F., & Karkar, A. (2016). Customer satisfaction factor in digital content marketing: Isparta Province as an example. *International Journal of Social Science Studies, 4,* 124.

Das, S. (2020). Innovations in digital banking service brand equity and millennial consumerism. In *Digital transformation and innovative services for business and learning* (pp. 62–79). IGI Global. https://doi.org/10.4018/978-1-7998-5175-2.ch004.

Das, S. (2021). *Search engine optimization and marketing: A recipe for success in digital marketing* (1st ed.). CRC Press. https://doi.org/10.1201/9780429298509.

Das, S., & Mondal, S. (2016). *Yours Digitally: Patanjali.* Available at SSRN 2737299.

Das, S., Mondal, S. R., Sahoo, K. K., Nayyar, A., & Musunuru, K. (2018). Study on Impact of socioeconomic make up of Facebook users on purchasing behavior. *Revista Espacios, 39,* 28–42. Retrieved from http://www.revistaespacios.com/a18v39n33/18393328.html.

Das, S., & Nayyar, A. (2019). Innovative ideas to manage urban traffic congestion in cognitive cities. In *Driving the development, management, and sustainability of cognitive cities* (pp. 139–162). IGI Global. https://doi.org/10.4018/978-1-5225-8085-0.ch006.

Das, S., & Nayyar, A. (2020). Effect of consumer green behavior perspective on green unwavering across various retail configurations. In *Green marketing as a positive driver toward business sustainability* (pp. 96–124). IGI Global. https://doi.org/10.4018/978-1-5225-9558-8.ch005.

Das, S., Nayyar, A., & Singh, I. (2019). An assessment of forerunners for customer loyalty in the selected financial sector by SEM approach toward their effect on business. *Data Technologies and Applications, 53*(4), 546–561. https://doi.org/10.1108/DTA-04-2019-0059.

Duy, N. T., Mondal, S. R., Van, N. T. T., Dzung, P. T., Minh, D. X. H., & Das, S. (2020). A study on the role of web 4.0 and 5.0 in the sustainable tourism ecosystem of Ho Chi Minh City, Vietnam. *Sustainability, 12*(17), 7140. https://doi.org/10.3390/su12177140.

Govindarajan, V., & Srivastava, A. (2020, March, 31). What the shift to virtual learning could mean for the future of higher ed. *Harvard Business Review*.

Gupta, D. K., Jena, D., Samantaray, A. K., & Das, S. (2019). HRD climate in selected public sector banks in India. *Revista ESPACIOS*, *40*(11).

Jain, S., Jain V., & Das, S. (2018). Relationship analysis between emotional intelligence and service quality with special evidences from Indian banking sector. *Revista ESPACIOS*, *39*(33).

Luna-Nevarez, C., & McGovern, E. (2018). On the use of mobile apps in education: The impact of digital magazines on student learning. *Journal of Educational Technology Systems, 47*(1), 17–31.

Mohanty, P. C., Dash, M., Dash, M., & Das, S. (2019). A study on factors influencing training effectiveness. *Revista Espacios, 40,* 7–15. Retrieved from http://www.revistaespacios.com/a19v40n02/19400207.html.

Mondal, S. R. (2020). A systematic study for digital innovation in management education: An integrated approach towards problem-based learning in Vietnam. In *Digital Innovations for Customer Engagement, Management, and Organizational Improvement* (pp. 104–120). IGI Global. https://doi.org/10.4018/978-1-7998-5171-4.ch006.

Mondal, S. R., Das, S., Musunuru, K., & Dash, M. (2017). Study on the factors affecting customer purchase activity in retail stores by confirmatory factor analysis. *Revista Espacios*, *38*(61).

Mondal, S. R., Mall, M., Mishra, U. S., & Sahoo, K. (2017). Investigating the factors affecting customer purchase activity in retail stores. *Revista Espacios*, *38*(57).

Mondal, S. R., & Sahoo, K. K. (2020). A study of green building prospects on sustainable management decision making. In *Green building management and smart automation* (pp. 220–234). IGI Global.https://doi.org/10.4018/978-1-5225-9754-4.ch011.

Nadanyiova, M., & Das, S. (2020). Millennials as a target segment of socially responsible communication within the business strategy. *Littera Scripta, 13*(1), 119–134. https://doi.org/10.36708/Littera_Scripta2020/1/8.

Sharma, E., & Das, S. (2020). Measuring impact of Indian ports on environment and effectiveness of remedial measures towards environmental pollution.

International Journal Environment and Waste Management, 25(3), 356–380. https://doi.org/10.1504/IJEWM.2019.10021787.

Singh, I., Nayyar, A., & Das, S. (2019). A study of antecedents of customer loyalty in banking & insurance sector and their impact on business performance. *Revista ESPACIOS, 40*(6).

Singh, I., Nayyar, A., Le, D.H., & Das, S. (2019). A conceptual analysis of internet banking users' perceptions. An Indian perceptive. *Revista ESPACIOS, 40*(14).

Singh, L. B., Mondal, S. R., & Das, S. (2020). Human resource practices & their observed significance for Indian SMEs. *Revista ESPACIOS, 41*(07). Retrieved from http://www.revistaespacios.com/a20v41n07/20410715.html.

Singh, S., & Das, S. (2018). Impact of post-merger and acquisition activities on the financial performance of banks: a study of Indian private sector and public sector banks. *Revista Espacios Magazine, 39*(26), 25.

Singh, S., Mondal, S., Singh, L. B., Sahoo, K. K., & Das, S. (2020b). An empirical evidence study of consumer perception and socioeconomic profiles for digital stores in Vietnam. *Sustainability, 12*(5), 1716.

Syrjämäki, A. H., Isokoski, P., Surakka, V., Pasanen, T. P., & Hietanen, J. K. (2020). Eye contact in virtual reality—A psychophysiological study *Computers in Human Behavior, 112,* 106454.

Van, N. T. T., Vrana, V., Duy, N. T., Minh, D. X. H., Dzung, P. T., Mondal, S. R., & Das, S. (2020). The role of human–machine interactive devices for post-COVID-19 innovative sustainable tourism in Ho Chi Minh City, Vietnam. *Sustainability, 12*(22), 9523. https://doi.org/10.3390/su12229523.

Venkatesh, V., Thong, J. Y., & Xu, X. (2012). Consumer acceptance and use of information technology: extending the unified theory of acceptance and use of technology. *MIS Quarterly, 36*(1), 157–178.

Internet Sources

Crawford, S. (2017). *Thinking Beyond the Cloud with Digital Applications.* https://blogs.gartner.com/steve-crawford/2017/03/21/thinking-beyond-the-cloud-with-digital-applications.

Digital and Social Media Landscape in India. (2020). https://sannams4.com/digital-and-social-media-landscape-in-india/.

KPMG (India) and Google. (2017). *Indian Languages-Defining India's Internet.* https://home.kpmg/in/en/home/insights/2017/04/indian-language-internet-users.html.

Lundberg, A., & Westerman, G. (2020). The Transformer CLO. *Harvard Business Review.* January–February 2020 Issue.

Nitecki, S. (2019). *Companies that Nailed Digital Transformation with Innovative Mobile Apps.* https://www.netguru.com/blog/innovative-mobile-enterprise-apps.

CHAPTER 7

Digital Entertainment Based Do It Yourself Content & Advertisement as a Factor of Driving Force for Trust & Customer Patronage

S. Ravi and Subhra R. Mondal

1 Introduction

Nowadays, digital marketing is too important for companies. It is mainly adopted among the business processes. It has the potential of increasing sales and revenues, as well as is important for building brand trust and loyalty (Chaffey & Ellis-Chadwick, 2016; Nadanyiova & Das, 2020). Digital marketing mention the value of goods, products, or services to customers, on the Internet (Behera, Gunasekaran, Gupta, Kamboj, & Bala, 2019; Singh, Mondal, Singh, Sahoo, & Das, 2020). In simple way, we say that Digital Marketing uses ICT to achieve marketing objectives.

S. Ravi
Faculty of Communication, Saint Leo University, Savannah, GA, USA

S. R. Mondal (✉)
The Honors Programme, Duy Tan University, Da Nang, Vietnam
e-mail: subhraranimondal@duytan.edu.vn

S. Das and S. Gochhait (eds.), *Digital Entertainment*,
https://doi.org/10.1007/978-981-15-9724-4_7

The three main uses of the Internet (Busca & Bertrandias, 2020; Duy et al., 2020; Sabogal Salamanca, Fajardo-Toro, Renteria, Mayett Moreno, & Sanchez Baltasar, 2020; Singh, Mondal, & Das, 2020; Van et al., 2020):

- To promote Online Advertising.
- To deliver e-Commerce services.
- To build strong bonding with customers, known as Social Media Marketing

A continuing relationship with the customers is important for E-business. Customers buy things to satisfy their needs and has a unique buying pattern (Wagner, Schramm-Klein, & Steinmann, 2018). Online shopping increases the value of customers and builds sustainable capabilities. And that's human nature which makes consumers to buy items recommended by their trustable people (Luo, Baker, & Donthu, 2019; Sharma & Das, 2020). The ways consumer know about the products or services are a special concern for marketing. One of these ways is the Consumer Word of Mouth (WOM), which help them to show the preferences and recommendations according to their experiences (Das, 2020; Luo, Baker, & Donthu, 2019; Strandberg, Styvén, & Hultman, 2019). With the increase of social media platforms and e-commerce sites, online WOM has become a main source of information (Chen & Yuan, 2020; Das, 2021; Mondal, 2020).

Online WOM is defined as electronic Word of Mouth, which by definition is "any positive or negative statement made by potential, actual, or former customers about a product or company, which is made available to a multitude of people and institutions via the Internet" (Das & Nayyar, 2020; Nam, Baker, Ahmad, & Goo, 2019). The content generated by users and consumers as online reviews and ratings, is a key features derived from the web 2.0, being important for E-Commerce and to generate sales (Laudon & Traver, 2019; Mondal & Sahoo, 2020). E-WoM comprises this content generated by any user in its different forms. The different mechanisms used by E-WoM to generate effects on sales have been analyzed through its effects on consumer´s decision making. One of the branches of analysis is related with the effects of E-WoM in generating brand trust, that could be later derived in purchase intention

(Das, Nayyar, & Singh, 2019; Ho-Dac, Carson, & Moore, 2013; Kostyra, Reiner, Natter, & Klapper, 2016; PowerReview, 2015).

Despite the fact there is abundant literature on E-WoM and its general effects on consumer behavior (see for example, Das & Nayyar, 2019; King, Racherla, & Bush, 2014b), is considered important to discuss in a more detailed way the effects of consumer-generated content, namely, ratings and reviews, on online purchasing behavior (Singh, Nayyar, Le, & Das, 2019). The chapter, will get an understanding on: (1) the types of consumer generated content and its influence in purchasing behavior online; (2) E-WoM as a source of trust; and (3) how E-WoM leverages brands with little recognition (Mohanty, Dash, Dash, & Das, 2019). In order to do that, the chapter is organized in the following parts besides this introduction: a background, which includes the main definitions related with E-WoM, its effects in consumer´s processing of information, in valence, and on willingness to recommend (Singh, Nayyar & Das, 2019). E-WoM is later revised in its different platforms such as social media and individual pages or blogs (Gupta, Jena, Samantaray, & Das, 2019). Online ratings and reviews and its helpfulness is discussed in a following chapter, and the effects of the consumer generated content on sales is revised (Singh & Das, 2018). The later parts summarize the findings with the main solutions and recommendations, future research directions and main conclusions (Jain, Jain, & Das, 2018).

2 Literature Review

2.1 E-WoM Main Definitions

Approximately 80% of Americans were online shoppers in 2016. About 74% of them think that it's important to read reviews posted by consumers who have purchased the items in which they have some interest to purchase. E-WoM has a significant impact on purchase intentions, product choice, product sales, and new product introductions (Craciun, Zhou, & Shan, 2019; Das, Mondal, Sahoo, Nayyar, & Musunuru, 2018; Mondal, Das et al., 2017). Consumers express their satisfaction or dissatisfaction about the products and services as they use in day to day through electronic means, which has been widely documented and is grouped in the academic literature as E-WoM (Das & Mondal, 2016; Mondal, Mall et al., 2017). Consumer´s opinions can have the following main forms:

- Online Rating
- Online Reviews
- Post or Comments in Social Media

2.2 Blogs

According to Gottschalk and Mafael (2017), sites where consumers find online ratings and reviews such as TripAdvisor.com and Amazon.com, complement these review texts with pieces of summary statistics, ratings, or information from the author. TripAdvisor.com, for example, allow customers to segment reviews according to the type of traveler: business, couple without children, family with children, and alike. Online reviews and ratings forums additionally provide recurrence distributions of ratings in the form of histograms which adds even more information (Guan & Lam, 2019). Yang, Hu, and Zhang (2007) verified that C2C markets (from consumer to consumer) achieved a better performance with the implementation of feedback systems such as reviews and ratings: in this way, buyers are more willing to trade and there is also an incentive for honest behavior from sellers, who see their profits increase while profits of dishonest behavior are reduced. George, Dahlander, Graffin, and Sim (2016) realized how online ratings work as an important signals, influence cognition of the perceived value of products and services to varying degrees for organizations and customers, somehow in the way that brands influence consumer behavior.

Mukhopadhyay and Chung (2016) argue that online rating environment can be described using three key aggregated measures: (1) valence or average rating, (2) volume or total number of ratings, and (3) variance or dispersion of ratings. Nevertheless, there are conflicting findings regarding extent and direction of such influences, which is caused by different levels of informational complexity given by these measures, requiring varying levels of cognitive effort in terms of processing and comprehension (Mukhopadhyay & Chung, 2016).

2.3 E-WoM and Processing of Information

As has been mentioned before, WoM refers to communications between consumers products or services that are considered to be more influential than sources from the marketplace (Lee & Youn, 2009). Similarly, Hennig-Thurau, Gwinner, Walsh, and Gremler (2004) defined E-WoM as "any positive or negative statement made by potential, actual, or former customers about a product or company, which is made available to a multitude of people and institutions via the Internet". Consumers give different value to E-WoM depending on where they found it. For instance, when someone finds a product review in a blog is more likely to review the characteristics and is less recommend the product than the review appears in the brand's website (Lee & Youn, 2009). Therefore, advertisers are careful with the practice of compensating consumers for product reviews on personal blogs. These practices can generate suspicions among consumers about positive product reviews on these type of blogs (Lee & Youn, 2009).

Krishnamurthy and Kumar (2018) conducted a study where taking consumers of India and the US, doing simulations about different products and services, to evaluate the brand preferences and expectations. The study suggests that high- versus low-involvement consumers will go more through E-WoM information and spend more time with them to develop an expectation or idea of the brand by creating a better image in their minds. During the search process, consumers go through hundreds of reviews for any single product or service (product reviews) as well as other information clues such as product's star ratings and author profiles so they use several other resources in order to skim the most valuable information (Gottschalk & Mafael, 2017). The following table presents some examples of these resources as different types of E-WoM information clues.

2.4 Valence of E-WoM

As has been explained above, the eWoM can be positive or negative, but that comments and evaluations have an influence on the firm or brand reputation. Nisar, Prabhakar, Ilavarasan, and Baabdullah (2019), test the hypothesis that both information diversity and valence increase a firm's financial performance, as well as enhances a firm's reputation.

According to Sen and Lerman (2007), people have a tendency to give a bigger weight to negative information compared to positive, but in case of utilitarian versus hedonic products, consumers reveal a negative bias for utilitarian product reviews only. Research has found that a negative E-WoM has a stronger effect that of positive E-WoM, the negativity bias. Furthermore, the inclusion of positive and negative E-WoM communication has a positive influence in the E-WoM credibility (Cheung & Thadani, 2012).

2.5 Effects of E-WoM in Consumer's Willingness to Recommend

E-WOM's ability to speak in a honest way about a product or store is beneficial for the consumer because it diminish problems that can arise from a lack of information, such as being able to see only positive online reviews and not the negative ones (King, Racherla, & Bush, 2014b). According to Mukhopadhyay and Chung (2016), people have higher disposition to recommend a product when the review is positive than when it is negative regardless of the E-WoM platform. This is because consumers are easily exposed to a large positive clues than negative ones, hence negative clues are perceived as counter intuitive (Sen & Lerman, 2007). E-WoM credibility in that sense is key for adoption, and is mediated by usefulness. Both, usefulness and credibility are associated with purchase intention (Cheung & Thadani, 2012).

2.6 E-WoM in Social Media

The research of E-WoM in social media has a great importance for the management of modern enterprises. J. Wang, Wang, and Li (2019) elaborate a discrete model to evaluate the effect of negative and positive E-WoM, finds that in OSN, enterprises choose more users to disseminate positive opinions and have control over negative views which reduce the economic losses. The source reliability, expertise, and knowledge has also been found important for social media for Sotiriadis and van Zyl (2013).

How close and important the consumer thinks about the source of information in a social media context is going to have a notable influence on the decision to search or pass other´s opinions on social media sites. In fact, this behavior tends to occur within a diverse and varied network of connections, and not necessarily only with close ties (close friends) or with persons with similar psychographic profiles (Chu & Kim, 2011). The degree of involvement is also important in these sites, in the case of high involvement services like tourism, consumers are going to devote more time to search other sources such as travel agencies, blogs and the implication for marketers is that monitoring all the comments in the social media platform for responding appropriately is even more important than posting (Sotiriadis & van Zyl, 2013).

2.7 E-WoM in Individual Pages or Blogs

Most comparisons on the effect on sales of E-WoM have been conducted in low-involvement categories. When conducted for a high involvement category such as digital cameras, studies have found that there is a difference between the influence in sales depending on the type of E-WoM site. If is hosted by the company, the effect is limited, while if the site is independent or external, the effect on sales is significant (Gu, Park, & Konana, 2012). Consumer credibility and usefulness of the E-WoM information is going to be moderated by the level of involvement, and consumers facing a high involvement decision are going to put more cognitive effort and use the central route of information processing (Cheung & Thadani, 2012).

There are some product or service categories where E-WoM has to achieve success, due to the existence of online communities for exchanging opinions about product and services. Thus, making it possible to create and adapt different analytic techniques to inspect the effects of reviews on sales of the products (Dellarocas, Zhang, & Awad, 2007).

2.8 Online Ratings

Online ratings help consumers to understand how they enjoy a product, with high quality (M. Sun, 2012). Willems, Waldner, and Ronquillo (2019), argue that online ratings are complements of information rather than its substitutes when the rating is represented with the image of stars which attracts more attention. As if people noticed easier that a product has 3 stars (2 are missing) than when it has 5 (they are complete). Similarly, Li, Wu, and Mai (2019), explain that the online rating is a quantitative summary of the reviewer's opinions or feelings toward a product or service, usually expressed as number of stars. Duan, Gu, and Whinston (2008) contrarily argue that it is not higher ratings what drives sales and suggest that "consumers are not influenced by the persuasive effect of online word-of-mouth, although they are affected by awareness effect generated by the underlying process of word-of-mouth" then, as a recommendation for business practitioners, they suggest that marketers should encourage consumers to use different means of dispersion of E-WoM rather than promoting or influence online ratings. B. Li et al. (2016) also found a correlation between sales, the volume of reviews and online ratings, however, their finding suggest that the influence of the latter two is stronger in search products than in experience products.

Anderson and Magruder (2012) find that high ratings cause restaurants using Yelp platform to sell out the prime Table 19% more frequently. Interestingly, the percentage increases to 27% in restaurants where information is more scarce (not externally accredited for example). Variance in the product ratings is also important, and high variance is associated with a niche product. Only in the case where the average rating is low and the variance is high, there will be a subsequent effect in demand (M. Sun, 2012).

2.9 Online Reviews

Mudambi and Schuff (2010) describe customer reviews as information that supplements "information provided by electronic storefronts such as product descriptions, reviews from experts, and personalized advise" are generated by automated recommendation systems. Park (2018) defines online reviews as those comments that a consumer shares through electronic means about the experience he or she has had with a product or service and that is shared in the hope that it will serve others to make a better informed decision at the moment of the purchase. Some consider that online reviews provide an authentic and independent evaluation from the perspective of the consumer and influences the attitudes of consumers toward a brand (Global Web Index, 2018).

The fast development of information technologies and the Internet have caused consumers to increasingly buy products and services and publish comments related to them through websites (Bi, Liu, Fan, & Zhang, 2019). Only on Tripadvisor.com it is estimated that more than 660 million reviews are available (Global Web Index, 2018). Online reviews are the third source of information about a product, the first source is search engines and the second is social networks (Global Web Index, 2018). As the reviews are important for sales, the identities and personal information of the reviewers are also very important.

2.10 Helpfulness of Online Reviews

Doing transactions between business and consumers, provide information based on the personal experience posted on the web by previous buyers of a specific product, turning it as free "sales assistants" to help more consumers on the buying decision of a product of interest (Fresneda & Gefen, 2019). For this reason, these days, many e-commerce enable and encourage customers to review the post in both the buying process as well the product quality, helping to reduce uncertainties around the experience (X. Sun, Han, & Feng, 2019). Srivastava and Kalro (2019) conducted a study, taking a series of reviews with different characteristics, and with the application of causal statistical tools, finds that when a review is comprehensive and complete, it provides inputs (Bickart and Schindler, 2001). On the other hand, when a review is readable and has clarity in content and presentation. Finally, they established that the effect of valence is complex and leads to different perceptions depending on the context.

According to Mudambi and Schuff (2010), the perception of helpfulness of an online review can be affected by three different factors: (1) extremity of review, (2) review depth, and (3) product type (experience vs. search goods). For instance, if the review has associated an extreme rating, then it is perceived as less helpful than those with moderated ratings for experience goods, however, this effect has not been found in search goods. On the other hand, when the commenter expresses extreme points of views, their reviews seem less credible. When dealing with the length of the review, the authors found that it "increases the diagnostic of a search good review more than that of an experience good review" and that "additional length in the textual review cannot compensate or substitute for sampling" in experience goods.

3 Effects of Consumer Generated Content on Sales

According to Li and Hitt (2008), online reviews have a considerable impact on sales of non-durable products as it is in these kind of products where customers cannot do quality control by testing all alternatives. Their research empirically showed that if early buyers of a product liked it, the company has an incentive to increase prices and take advantage of self-selection. In the same way, if early buyers are critical in relation to the product, the company could reduce the price by seeking to balance early losses in exchange for future profits. They claim that consumer heterogeneity increases profit only if they buy in advance are "advocates" of the product; otherwise, the heterogeneity of the consumer affects negatively because they are willing to pay a high price given access to perfect information are deterred by negative criticism (Li & Hitt, 2008). It is noteworthy from the study of Chevalier and Mayzlin (2006) that customers read all reviews and it is with this complete information that they make their decision, however, Guan and Lam (2019) showed that consumers only read a large number of reviews when the online rating (given by the average stars) does not confirm their expectations.

According to Wang, Zhang, and Zheng (2011), B2C e-commerce prices directly affect the perceived profit and comments made after purchase, which determines the volume of sales and the benefit of the products. The impact of online reviews on prices is very important for electronic retailers to set reasonable price and increase profits (Erkan and Evans, 2016). Li and Hitt (2008) also highlight the value of honesty in advertising because with the rise of online reviews, if advertising creates false expectations, disappointed early consumer reviews will end up hurting the company's profits.

Li et al. (2019) argue that the information included in the review drives sales and helps companies to exploit the value of E-WoM by promoting new products, attracting and retaining buyers. However, this is a combined effect between ratings and online reviews, highlighting that the influence of sales is completely mediated by star rating (Forman, Ghose, & Wiesenfeld, 2008). Sen and Lerman (2007), concluded that consumer reviews consideration in purchasing decision making is mediated by the type of product. In this regard, they found that consumers are likely to consider the consumer reviews when trying to buy hedonic products than with utilitarian products.

4 Solutions and Recommendations

The WOM is one of the most important ways that can influence positively or negatively companies and brands (Hu, Koh, & Reddy, 2014). However, based on the fact of the great penetration that the Internet has, implies that all companies and business have to include the online business on their marketing strategies, using the net as a base to connect with their clients (Hu, Liu, & Zhang, 2008). That obliges to having to resort to the use of strategies associated with digital marketing, either to advertise or to do direct sales of products and services. Under the axiom that the customers and consumers always will have an opinion about their experience, one characteristic that has the online environment is the possibility of the registry of those opinions. This characteristic could be managed as an advantage, which implies that is necessary to implement mechanisms and tools that allow registering and manage all the possible information that can be given by the customer.

To achieve this aim, it is recommended to implement strategies that encourage the customers to write their opinions or reviews, as well as the construction of data warehouses and data mining models that allow extracting the implicit knowledge on the information produce on those reviews.

Likewise, E-WoM as a source of trust must be managed, meaning that it is hard for brands to keep products and services performing at optimal levels all the time. Therefore, if criticism arises at some point, marketing practitioners should try to engage in direct contact with the consumer and, when possible, offer public solution. In this way, other consumers will notice that the brand cares and might trust more in the product or service than when the criticism is left alone. Consumers complement the E-WoM information found in the E-Commerce place with other information from other sources (Gottschalk & Mafael, 2017). This implies that the marketers should be able to track the different customer journey paths followed by the potential buyer and adapt the communication to the different stages of the purchase decision. Remarketing optimized techniques, a good and informative E-Commerce place and alliances with other content providers meaningful for the customer would be of great help.

Marketers must be able to track the customers in each step of the purchase for that matter, and also to have specific metrics in each step of the sales funnel relating the E-WoM with other measures (King, Racherla, & Bush, 2014a; Moe, Trusov, & Smith, 2011). Monitoring products and categories, their evolution and its relationship with E-WoM content generated by consumers can help to react early and rapidly to potential problems related with the product information or the product performance, as well as adapt the marketing strategies depending on the metrics found (Kumar & Benbasat, 2006; Yong, 2006).

5 Future Exploration of the Topic

There are some research lines, but two have a special interest which are: first, using big data and data mining techniques to explore the behavior of consumers on different retails webs, establishing the more relevant variables; second, but combined with the latter, is to compare consumer of different geographical zones and define if there are differences on the consumer behavior according to that. An interesting line of research will be related with the specific emotions and reviews. Previous works have showed the importance of negative emotions and perceived utility of online reviews (Ren & Hong, 2019). Nevertheless, more studies could shed light on the different effects of different emotions, positive and negative, in the different phases of the purchase process, given the fact that emotions are very important for a specific product choice.

6 Conclusion

Good E-WoM has the potential to leverage trust in the same way that branding does. Brands give consumers the assurance they need prior to purchase the product or service according to their expectations. In that sense, E-WoM comes as a way to assure quality too. Major brands need to rely on advertising to build their recognition, but also comply their promises trough good quality of products and services, which in turn, will give them good E-WoM. Thus, creating a virtuous circle that in the end will give more value for the brand. On the other hand, products not highly branded rely exclusively on E-WoM to build trust.

When deciding a purchase consumers have access to many types of E-WoM. Not all of them are considered for every online purchase nor for every kind of product or service. However, they all provide different clues of information and altogether they increase the trust in the product. E-WoM could be found on the website of a retailer or manufacturer or in a more independent form, meaning that the provider or the dealer does not have control over what it is said about their products or services. For instance, when consumers post their opinion on social media, a search engine, or a personal blog. Either way, the main types of E-WoM are (1) online reviews, (2) online ratings, and E-WoM on social media and independent blogs.

Online ratings are mostly presented on a scale of zero to five stars and are usually accompanied with the number of people who have assessed the product or service. Similarly, some retailers provide means to give a star rating to not only the overall experience with the product or service, but to different aspects of the buying and consumption experience, for instance, the shipping and handling.

Online ratings are rarely evaluated in isolation. They are used in conjunction with online reviews as means of confirmation of one's previous expectations with the brand or the product. For instance, if the product is highly branded but has low star ratings, consumers might try to find an explanation on the reviews. Likewise, when the product is off-branded and has high star ratings. Similarly, the number of reviews are also considered as clues prior to purchase.

Consumers weight higher the E-WoM in social media and independent blogs than E-WoM in manufacturer's or retailer's websites. This is because most consumers have doubts on whether manufacturers or retailers' moderates consumers' feedback allowing only good posts and ratings to show to others. However, this product type has a mediating effect and the impact of E-WoM on sales. Products described as hedonic or experience goods get a greater impact than utilitarian or search goods. This is because when consumers are buying search goods they can assess quality or features more easily than for experience goods or services.

References

Anderson, M., & Magruder, J. (2012). Learning from the crowd: Regression discontinuity estimates of the effects of an online review database. *Economic Journal, 122*(563), 957–989. https://doi.org/10.1111/j.1468-0297.2012.02512.x.

Behera, R. K., Gunasekaran, A., Gupta, S., Kamboj, S., & Bala, P. K. (2019, March). Personalized digital marketing recommender engine. *Journal of Retailing and Consumer Services*, 101799. https://doi.org/10.1016/j.jretconser.2019.03.026.

Bi, J.-W., Liu, Y., Fan, Z.-P., & Zhang, J. (2019). Wisdom of crowds: Conducting importance-performance analysis (IPA) through online reviews. *Tourism Management, 70,* 460–478. https://doi.org/10.1016/j.tourman.2018.09.010.

Bickart, B., & Schindler, R. M. (2001). Internet forums as influential. *Journal of Interactive Marketing, 15*(3), 31–40.

Busca, L., & Bertrandias, L. (2020). A framework for digital marketing research: Investigating the four cultural eras of digital marketing. *Journal of Interactive Marketing, 49,* 1–19. https://doi.org/10.1016/j.intmar.2019.08.002.

Chaffey, D., & Ellis-Chadwick, F. (2016). *Digital Marketing* (6th ed.). Edinburgh: Pearson.

Chen, Z., & Yuan, M. (2020). Psychology of word of mouth marketing. *Current Opinion in Psychology, 31,* 7–10. https://doi.org/10.1016/j.copsyc.2019.06.026.

Cheung, C. M. K., & Thadani, D. R. (2012). The impact of electronic word-of-mouth communication: A literature analysis and integrative model.

Decision Support Systems, 54(1), 461–470. https://doi.org/10.1016/j.dss.2012.06.008.

Chevalier, J. A., & Mayzlin, D. (2006). The effect of word of mouth on sales: Online book reviews. *Journal of Marketing Research, 43*(3), 345–354. https://doi.org/10.1509/jmkr.43.3.345.

Chu, S.-C. C., & Kim, Y. (2011). Determinants of consumer engagement in electronic Word-Of-Mouth (E-WoM) in social networking sites. *International Journal of Advertising, 30*(1), 47–75. https://doi.org/10.2501/IJA-30-1-047-075.

Craciun, G., Zhou, W., & Shan, Z. (2019, January). Discrete emotions effects on electronic word-of-mouth helpfulness: The moderating role of reviewer gender and contextual emotional tone. *Decision Support Systems, 130,* 113226. https://doi.org/10.1016/j.dss.2019.113226.

Das, S. (2020). Innovations in digital banking service brand equity and millennial consumerism. In *Digital transformation and innovative services for business and learning* (pp. 62–79). IGI Global. https://doi.org/10.4018/978-1-7998-5175-2.ch004.

Das, S. (2021). *Search engine optimization and marketing: A recipe for success in digital marketing* (1st ed.). CRC Press. https://doi.org/10.1201/9780429298509.

Das, S., Mondal, S. R., Sahoo, K. K., Nayyar, A., & Musunuru, K. (2018). Study on Impact of socioeconomic make up of Facebook users on purchasing behavior. *Revista Espacios, 39,* 28–42. Retrieved from http://www.revistaespacios.com/a18v39n33/18393328.html.

Das, S., & Mondal, S. (2016). *Yours Digitally: Patanjali.* Available at SSRN 2737299.

Das, S., & Nayyar, A. (2020). Effect of consumer green behavior perspective on green unwavering across various retail configurations. In *Green marketing as a positive driver toward business sustainability* (pp. 96–124). IGI Global. https://doi.org/10.4018/978-1-5225-9558-8.ch005.

Das, S., Nayyar, A., & Singh, I. (2019). An assessment of forerunners for customer loyalty in the selected financial sector by SEM approach toward their effect on business. *Data Technologies and Applications, 53*(4), 546–561. https://doi.org/10.1108/DTA-04-2019-0059.

Das, S., & Nayyar, A. (2019). Innovative Ideas to Manage Urban Traffic Congestion in Cognitive Cities. In *Driving the Development, Management, and Sustainability of Cognitive Cities* (pp. 139–162). IGI Global. https://doi.org/10.4018/978-1-5225-8085-0.ch006.

Dellarocas, C., Zhang, X., & Awad, N. F. (2007). Exploring the value of online product reviews in forecasting sales: The case of motion pictures. *Journal of Interactive Marketing, 21*(4), 23–45. https://doi.org/10.1002/dir.20087.

Duan, W., Gu, B., & Whinston, A. B. (2008). Do online reviews matter? An empirical investigation of panel data. *Decision Support Systems, 45*(4), 1007–1016. https://doi.org/10.1016/j.dss.2008.04.001.

Duy, N. T., Mondal, S. R., Van, N. T. T., Dzung, P. T., Minh, D. X. H., & Das, S. (2020). A study on the role of web 4.0 and 5.0 in the sustainable tourism ecosystem of Ho Chi Minh City, Vietnam. *Sustainability, 12*(17), 7140. https://doi.org/10.3390/su12177140.

Erkan, I., & Evans, C. (2016). The influence of E-WoM in social media on consumers' purchase intentions: An extended approach to information adoption. *Computers in Human Behavior*. https://doi.org/10.1016/j.chb.2016.03.003.

Forman, C., Ghose, A., & Wiesenfeld, B. (2008). Examining the relationship between reviews and sales: The role of reviewer identity disclosure in electronic markets. *Information Systems Research, 19*(3), 291–313. https://doi.org/10.1287/isre.1080.0193.

Fresneda, J. E., & Gefen, D. (2019). A semantic measure of online review helpfulness and the importance of message entropy. *Decision Support Systems, 125*(July), 113117. https://doi.org/10.1016/j.dss.2019.113117.

George, G., Dahlander, L., Graffin, S., & Sim, S. (2016). From the editors reputation and status: Expanding the role of social evaluations. *Academy of Management Journal, 59*(1), 1–13. https://doi.org/10.5465/amj.2016.4001.

Global Web Index. (2018). *Product Reviews and Commerce. Examining the role of product reviews in the digital purchase journey.* Retrieved from https://www.globalwebindex.com/hubfs/Downloads/Product-Reviews-and-Commerce-report.pdf.

Gottschalk, S. A., & Mafael, A. (2017). Cutting through the online review jungle—Investigating selective E-WoM processing. *Journal of Interactive Marketing, 37,* 89–104. https://doi.org/10.1016/j.intmar.2016.06.001.

Gu, B., Park, J., & Konana, P. (2012). The impact of external word-of-mouth sources on retailer sales of high-involvement products. *Information Systems Research, 23*(1), 182–196. https://doi.org/10.1287/isre.1100.0343.

Guan, C., & Lam, S. Y. (2019). Product rating statistics as consumer search aids. *Journal of Interactive Marketing*. https://doi.org/10.1016/j.intmar.2019.02.003.

Gupta, D. K., Jena, D., Samantaray, A. K., & Das, S. (2019). HRD climate in selected public sector banks in India. *Revista ESPACIOS, 40*(11).

Hennig-Thurau, T., Gwinner, K. P., Walsh, G., & Gremler, D. D. (2004). Electronic word-of-mouth via consumer-opinion platforms: What motivates consumers to articulate themselves on the Internet? *Journal of Interactive Marketing, 18*(1), 38–52. https://doi.org/10.1002/dir.10073.

Ho-Dac, N. N., Carson, S. J., & Moore, W. L. (2013). The effects of positive and negative online customer reviews: Do brand strength and category maturity matter? *Journal of Marketing, 77*(6), 37–53. https://doi.org/10.1509/jm.11.0011.
Hu, N., Koh, N. S., & Reddy, S. K. (2014). Ratings lead you to the product, reviews help you clinch it? The mediating role of online review sentiments on product sales. *Decision Support Systems, 57*(1), 42–53. https://doi.org/10.1016/j.dss.2013.07.009.
Hu, N., Liu, L., & Zhang, J. J. (2008). Do online reviews affect product sales? The role of reviewer characteristics and temporal effects. *Information Technology and Management, 9*(3), 201–214. https://doi.org/10.1007/s10799-008-0041-2.
Jain, S., Jain V., & Das, S. (2018). Relationship analysis between emotional intelligence and service quality with special evidences from Indian banking sector. *Revista ESPACIOS, 39*(33).
King, R. A., Racherla, P., & Bush, V. D. (2014a). What we know and don't know about online word-of-mouth: A review and synthesis of the literature. *Journal of Interactive Marketing, 28*(3), 167–183. https://doi.org/10.1016/j.intmar.2014.02.001.
King, R. A., Racherla, P., & Bush, V. D. (2014b). What we know and don't know about online word-of-mouth: A review and synthesis of the literature. *Journal of Interactive Marketing*. https://doi.org/10.1016/j.intmar.2014.02.001.
Kostyra, D. S., Reiner, J., Natter, M., & Klapper, D. (2016). Decomposing the effects of online customer reviews on brand, price, and product attributes. *International Journal of Research in Marketing, 33*(1), 11–26. https://doi.org/10.1016/j.ijresmar.2014.12.004.
Krishnamurthy, A., & Kumar, S. R. (2018, February). Journal of Retailing and Consumer Services Electronic word-of-mouth and the brand image: Exploring the moderating role of involvement through a consumer expectations lens. *Journal of Retailing and Consumer Services, 43*, 149–156. https://doi.org/10.1016/j.jretconser.2018.03.010.
Kumar, N., & Benbasat, I. (2006). The influence of recommendations and consumer reviews on evaluations of websites. *Information Systems Research, 17*(4), 425–439. https://doi.org/10.1287/isre.1060.0107.
Laudon, K. C., & Traver, C. G. (2019). *E-Commerce 2019: Business, Technology and Society* (17th ed.). Edinburgh: Pearson.
Lee, M., & Youn, S. (2009). Electronic word of mouth (E-WoM): How E-WoM platforms influence consumer product judgement. *International Journal of Advertising, 28*(3), 473–499. https://doi.org/10.2501/S0265048709200709.
Li, B., Ch'ng, E., Chong, A. Y.-L. L., Bao, H., Ch'ng, E., Chong, A. Y.-L. L., … Bao, H. (2016). Predicting online e-marketplace sales performances: A big

data approach. *Computers & Industrial Engineering*, *101*, 565–571. https://doi.org/10.1016/j.cie.2016.08.009.

Li, Xiaolin, Wu, C., & Mai, F. (2019). The effect of online reviews on product sales: A joint sentiment-topic analysis. *Information & Management, 56*(2), 172–184. https://doi.org/10.1016/j.im.2018.04.007.

Li, Xinxin, & Hitt, L. M. (2008). Self-selection and information role of online product reviews. *Information Systems Research, 19*(4), 456–474. https://doi.org/10.1287/isre.1070.0154.

Luo, A., Baker, A., & Donthu, N. (2019, July). Capturing dynamics in the value for brand recommendations from word-of-mouth conversations. *Journal of Business Research, 104*, 247–260. https://doi.org/10.1016/j.jbusres.2019.07.015.

Moe, W. W., Trusov, M., & Smith, R. H. (2011). The value of social dynamics in online product ratings forums. *Journal of Marketing Research, 48*(3), 444–456. https://doi.org/10.1509/jmkr.48.3.444.

Mohanty, P.C., Dash, M., Dash, M., & Das, S. (2019). A study on factors influencing training effectiveness. *Revista Espacios, 40,* 7–15. Retrieved from http://www.revistaespacios.com/a19v40n02/19400207.html.

Mondal, S. R. (2020). A systematic study for digital innovation in management education: An integrated approach towards problem-based learning in Vietnam. In *Digital innovations for customer engagement, management, and organizational improvement* (pp. 104-120). IGI Global. https://doi.org/10.4018/978-1-7998-5171-4.ch006.

Mondal, S., & Sahoo, K. K. (2020). A study of green building prospects on sustainable management decision making. In *Green building management and smart automation* (pp. 220–234). IGI Global. https://doi.org/10.4018/978-1-5225-9754-4.ch011.

Mondal, S., Mall, M., Mishra, U. S., & Sahoo, K. (2017). Investigating the factors affecting customer purchase activity in retail stores. *Revista Espacios*, *38*(57).

Mondal, S., Das, S., Musunuru, K., & Dash, M. (2017). Study on the factors affecting customer purchase activity in retail stores by confirmatory factor analysis. *Revista Espacios*, *38*(61).

Mudambi, S. M., & Schuff, D. (2010). What makes a helpful online review? A study of customer reviews on amazon.com. *MIS Quarterly: Management Information Systems*, *34*(1), 185–200. https://doi.org/10.2307/20721420.

Mukhopadhyay, S., & Chung, T. S. (2016). Preference instability, consumption and online rating behavior. *International Journal of Research in Marketing, 33*(3), 624–638. https://doi.org/10.1016/j.ijresmar.2015.11.007.

Nadanyiova, M., & Das, S. (2020). Millennials as a Target Segment of Socially Responsible Communication within the Business Strategy. *Littera Scripta, 13*(1), 119–134. https://doi.org/10.36708/Littera_Scripta2020/1/8.

Nam, K., Baker, J., Ahmad, N., & Goo, J. (2019, October). Determinants of writing positive and negative electronic word-of-mouth: Empirical evidence for two types of expectation confirmation. *Decision Support Systems*, 113168. https://doi.org/10.1016/j.dss.2019.113168.

Nisar, T. M., Prabhakar, G., Ilavarasan, P. V., & Baabdullah, A. M. (2019, September). Journal of Retailing and Consumer Services Up the ante: Electronic word of mouth and its effects on firm reputation and performance. *Journal of Retailing and Consumer Services*, 101726. https://doi.org/10.1016/j.jretconser.2018.12.010.

Park, Y. J. (2018). Predicting the helpfulness of online customer reviews across different product types. *Sustainability (Switzerland)*, *10*(6). https://doi.org/10.3390/su10061735.

PowerReview. (2015). *From Reviews to Revenue, how star ratings and review content influence purchase*. Retrieved from https://www.powerreviews.com/wp-content/uploads/2016/04/Northwestern-Vol1.pdf.

Ren, G., & Hong, T. (2019). Examining the relationship between specific negative emotions and the perceived helpfulness of online reviews. *Information Processing and Management*. https://doi.org/10.1016/j.ipm.2018.04.003.

Sabogal Salamanca, M., Fajardo-Toro, C. H., Renteria, J. C., Mayett Moreno, Y., & Sanchez Baltasar, L. B. (2020). Online Purchase and Advertising in Latin America. In T. Semerádová (Ed.), *Impacts of Online Advertising on Business Performance* (pp. 104–137). https://doi.org/10.4018/978-1-7998-1618-8.ch005.

Sen, S., & Lerman, D. (2007). Why are you telling me this? An examination into negative consumer reviews on the web. *Journal of Interactive Marketing*, *21*(4), 76–94. https://doi.org/10.1002/dir.20090.

Sharma, E., & Das, S. (2020). Measuring impact of Indian ports on environment and effectiveness of remedial measures towards environmental pollution. *International Journal Environment and Waste Management*, *25*(3), 356–380. https://doi.org/10.1504/IJEWM.2019.10021787.

Singh, S., Mondal, S., Singh, L. B., Sahoo, K. K., & Das, S. (2020). An empirical evidence study of consumer perception and socioeconomic profiles for digital stores in Vietnam. *Sustainability*, *12*(5), 1716.

Singh, L. B., Mondal, S. R., & Das, S. (2020). Human resource practices & their observed significance for Indian SMEs. *Revista ESPACIOS*, *41*(07). Retrieved from http://www.revistaespacios.com/a20v41n07/20410715.html.

Singh, I., Nayyar, A., Le, D.H., & Das, S. (2019). A conceptual analysis of internet banking users' perceptions. An Indian perceptive. *Revista ESPACIOS*, *40*(14).

Singh, I., Nayyar, A., & Das, S. (2019). A study of antecedents of customer loyalty in banking & insurance sector and their impact on business performance. *Revista ESPACIOS*, *40*(6).

Singh, S., & Das, S. (2018). Impact of post-merger and acquisition activities on the financial performance of banks: A study of Indian private sector and public sector banks. *Revista Espacios Magazine, 39*(26), 25.

Sotiriadis, M. D., & van Zyl, C. (2013). Electronic word-of-mouth and online reviews in tourism services: The use of twitter by tourists. *Electronic Commerce Research, 13*(1), 103–124. https://doi.org/10.1007/s10660-013-9108-1.

Srivastava, V., & Kalro, A. D. (2019). Science direct enhancing the helpfulness of online consumer reviews: The role of latent (content) factors. *Journal of Interactive Marketing, 48,* 33–50. https://doi.org/10.1016/j.intmar.2018.12.003.

Strandberg, C., Styvén, M. E., & Hultman, M. (2019, January). Places in good graces: The role of emotional connections to a place on word-of-mouth. *Journal of Business Research*, 1. https://doi.org/10.1016/j.jbusres.2019.11.044.

Sun, M. (2012). How does the variance of product ratings matter? *Management Science, 58*(4), 696–707. https://doi.org/10.1287/mnsc.1110.1458.

Sun, X., Han, M., & Feng, J. (2019, January). Helpfulness of online reviews: Examining review informativeness and classification thresholds by search products and experience products. *Decision Support Systems, 124*, 113099. https://doi.org/10.1016/j.dss.2019.113099.

The World Bank. (n.d.). *Individuals using the Internet (% of population) | Data*. Retrieved from https://data.worldbank.org/indicator/it.net.user.zs.

Van, N. T. T., Vrana, V., Duy, N. T., Minh, D. X. H., Dzung, P. T., Mondal, S. R., & Das, S. (2020). The role of human–machine interactive devices for post-COVID-19 innovative sustainable tourism in Ho Chi Minh City, Vietnam. *Sustainability, 12*(22), 9523. https://doi.org/10.3390/su12229523.

Wagner, G., Schramm-Klein, H., & Steinmann, S. (2018, November). Online retailing across e-channels and e-channel touchpoints: Empirical studies of consumer behavior in the multichannel e-commerce environment. *Journal of Business Research, 107*, 256–270. https://doi.org/10.1016/j.jbusres.2018.10.048.

Wang, H., Zhang, W., & Zheng, L. (2011). Dynamic pricing in B2C based on online product reviews. *Procedia Engineering, 23,* 270–275. https://doi.org/10.1016/j.proeng.2011.11.2501.

Wang, J., Wang, X., & Li, Y. (2019). A discrete electronic word-of-mouth propagation model and its application in online social networks. *Physica a, 527,* 121172. https://doi.org/10.1016/j.physa.2019.121172.

Willems, J., Waldner, C. J., & Ronquillo, J. C. (2019). Reputation Star Society: Are star ratings consulted as substitute or complementary information? *Decision Support Systems, 124,* 1–11. https://doi.org/10.1016/j.dss.2019.113080.

Yang, J., Hu, X., & Zhang, H. (2007). Effects of a reputation feedback system on an online consumer-to-consumer auction market. *Decision Support Systems, 44*(1), 93–105. https://doi.org/10.1016/j.dss.2007.03.005.
Yong, L. (2006). Word of mouth for movies: Its dynamics and impact on box office revenue. *Journal of Marketing*, *70*(3), 74–89. Retrieved from http://search.ebscohost.com/login.aspx?direct=true&db=buh&AN=21895698&lang=zh-cn&site=ehost-live.

CHAPTER 8

The Dynamic Rise of Digital Brands' Market Mavens in Digital Entertainment: A Complete Know How for Curious Customers

Öğr. Üyesi Orhan Duman and Subhankar Das

1 Introduction

In this intense competitive environment, if businesses want to gain advantages over their competitors must choose the most effective communication channels for the messages they will send to the target consumers and use them correctly (Nadanyiova & Das, 2020; Singh, Mondal, Singh, Sahoo, & Das, 2020). One of these communication channels is WOM (word-of-mouth) communication. This communication is becoming more important for businesses today. These days many businesses has started to realize the significance of WOM communication

Öğr. Ü. O. Duman
Ömer Seyfettin Faculty of Applied Sciences, Media and Communication Department, Bandırma Onyedi Eylül University, Bandırma, Turkey
e-mail: oduman@bandirma.edu.tr

S. Das (✉)
The Honors Programme, Duy Tan University, Da Nang, Vietnam
e-mail: subhankardas@duytan.edu.vn

S. Das and S. Gochhait (eds.), *Digital Entertainment*,
https://doi.org/10.1007/978-981-15-9724-4_8

which is mostly created by opinion leaders, innovators, early adopters and market mavens (Singh, Mondal, & Das, 2020). Market mavens are those who have a general market knowledge of the product, are extroverted, perfectionist and are willing to discuss, share information with other consumers (Martens, 2010). Businesses are exploring in ways to influence personal communication channels by providing positive advice to target consumers about the products and services by their brands (Sharma & Das, 2020). In this period, where customer satisfaction and consequently customer loyalty becomes increasingly important, WOM communication and viral marketing have become important in customers' purchasing decisions (Das, 2020). WOM communication has become effective on consumers as social media and virtual world with technological developments and widespread internet environment (Mondal, 2020). In light of these developments, WOM communication has become more common as electronic WOM (eWOM) communication eWOM communication has become a very important element which is affecting such as consumers' pre-purchase decisions and post-purchase satisfaction, business image, promotional activities (Mondal & Sahoo, 2020). The concept of "personal impact" is explained by the fact that eWOM and market mavens have a stronger impact on consumers' purchasing decisions than other marketing communication tools such as advertising, personal sales, sales development efforts, public relations and direct marketing activities (Das & Nayyar, 2020). The concept of personal impact and WOM communication are interrelated. In modern marketing research, this issue has been mentioned after the 1940s (Buttle, 1998: 241). Personal impact is defined as the effect or change that occurs in a person's attitudes or behaviors as a result of communication with others (Das, Nayyar, & Singh, 2019). Market mavens influence consumers' purchasing decisions by using personal impact and also have a various information about product features and point of sale (Das & Nayyar, 2019). This consumer group follows new goods and many brands. Even if they do not have detailed information about the goods, they keep track of where the goods are best sold (Puspa & Rahardja, 2009: 25). While companies try to gain an advantage in competition, they are trying to turn the complexity of intense advertising and promotion that shocks their budgets into a more efficient process (Singh, Nayyar, Le, & Das, 2019). They achieve this process by focusing on the power of WOM communication together with increasing and widespread interpersonal communication methods such as short messages, blocks, social networking sites, etc., through developing technology (Chelminski & Coulter, 2007: 70). In this respect, market mavens and other reference groups have become an important consumer mass that should be reached for enterprises, especially in promotion policies in today's competitive environment (Duy et al., 2020; Mohanty, Dash, Dash, & Das, 2019; Van et al. 2020).

The purpose of this study is to inspect the effect of market mavens on consumer behavior through eWOM and to draw attention to the consumer-based brand value that can be obtained through viral communication activities on brands (Das, 2021; Singh, Nayyar, & Das, 2019).

2 Reference Groups

Consumers often make decisions under the influence of their environment (Gupta, Jena, Samantaray, & Das, 2019). In particular, the consultative groups referred to by the individual are highly effective in consumer behavior. Reference groups are generally; groups in which the person interacts, such as family and friends, and groups in which he/she does not but wishes to be interact (Kotler, 2002: 89). The person finds this group as an important source of data, whether they are in it or not (Singh & Das, 2018). The reference groups generally manifest themselves in two ways on consumer behavior. First, they model the group's preferences and behave in the same way. Second, the purchase decision is made directly by the group (Jain, Jain, & Das, 2018). However, the effect of these groups on the consumer is not limited in receiving information or making a purchase decision. Generally, it is stated that the reference groups are directed by the effects of providing information, adding value and providing benefit on the consumer behaviors. Furthermore, the effect of the group on consumer's behavior is influenced by two factors, the product/brand is a luxury or mandatory one, and the use of the product/brand in personal or in public. When the product is a luxury one and used in common areas, then the group has a strong influence on both product and its brand preference. Conversely, when it comes to personal use and mandatory product, the effect of reference group for brand and product is weak (Solomon, 2007: 381–382). The apparent and status-consuming people wants to give some messages to others by showing either their social image or assets (Das, Mondal, Sahoo, Nayyar, & Musunuru, 2018). In fact, the satisfaction of these consumptions depends on the liking and appreciation of other people or reference groups (Mondal, Das et al., 2017). Therefore, reference groups are leading in making such consumptions (Chen, 2002: 33). The self is often considered a subjective concept and is thought to be different from the social structure. However, there are close relationships between the self and the social structure. Studies conducted between cultures suggest that the social structure in which people are involved has an effect on self-formation (Mondal, Mall et al., 2017). In other words, the relationships and grouping tendencies of individuals are effective on their selves. For example, it is stated that western cultures are mostly personal-centered and eastern cultures are relationship-centered. Being self-centered is primarily expressed as thinking about themselves and their interests, and this is called an independent self. Relationship-centered is to take into account other people and their relationship with them and it is called dependent self (Escalas & Betmann, 2005: 380). Therefore, those with a dependent self are influenced by other people or reference groups in all their behaviors and decisions. Consumption patterns and product/brand preferences are also the guiding and preferences of these groups (Hamilton, 2003: 492). Consumers are seeking information or approval from reference groups on two subjects which are collecting information and/or the preferences of others about products and brands. People here apply information by gathering from reference groups or adopting their preferences (Bearden, Netemeyer, & Teel, 1989: 473). Consumers are influenced by their identities and reference groups while they are giving various meanings to brands. But individuals prefer brands that support themselves, they also try to find brand meanings that support their reference groups (Escalas & Bettman, 2005: 380). It is observed that there are three types of consumers that can be defined as reference groups in terms of enterprises in receiving a product or service and transferring information. These are innovators, opinion leaders and market mavens (Brancaleone & Gountas, 2007: 522). While traditional marketing communication is accepted as the number of people that can be reached as value, in today's marketing communication approach, the value of a group in the market is not related to the volume but the effect (Das & Mondal, 2016). In this type of market, the reference group significantly affects the rest of the market and makes this group of influencers valuable for the companies. The added value generated by these groups for businesses has made reference groups attractive for marketers. The tendency of these consumer groups to obtain information about market, products and services, to spread and share this information is an important value for brands (Clark & Goldsmith, 2005: 290). There are three important groups that influence consumers' purchasing decisions. These are market mavens, opinion leaders and innovators. Although these three concepts are related to each other, they contain differences (Ruvivo & Shoham, 2007). While there are fundamental differences in their level of knowledge and behavior, these three groups of consumers play a key role in spreading products or services among others (Fitzmaurice, 2011: 72).

2.1 Innovations and Innovators

Innovation is the addition of meaningful value to what is changed through a successful creative process (Assink, 2006: 217). Because of innovations, economic and social resources are shifting from inefficient areas to productive areas. Innovation is graded according to the difference level of innovation compared to its predecessor. The least innovation is classified as graded innovation, modular innovation, architectural innovation, system innovation, radical innovation according to the highest innovation from the insider (Slaughter, 1998: 229). Innovation as the opinion, application or object that the individual perceives as new (Rogers, 1976: 292). Innovation spreads across countries, sectors, social classes and individuals. Innovations also change during the deployment to meet new needs. Spread is the spread of innovation through communication over a period of time. Adoption reviews the steps before consumers accept new products. Consumer innovation is a concept related to the adoption of innovations (Dobre, Dragomir, & Preda, 2009: 19–20). While some consumers can easily adopt innovations and even look forward to the launch of new products; for some, innovation brings risks and uncertainties. Different people have different levels of adoption of innovations. Innovators are also called early adapters because they adapt easily to innovation (Goldsmith, Flynn, & Goldsmith, 2003). According to the time of adopting innovations, people are divided into five different groups (Robertson, 1967: 16). They are innovators, early adopters, early acting majority, late majority and ending ones. On an individual scale, all individuals adopt the things and ideas they perceive as new and are therefore somewhat innovative. In this respect, it gives the market the desire to consume natural dynamism of society. Therefore, it is the consumer's desire for innovation that determines the direction of the market. The first step in the proper marketing of innovation is the identification and understanding of innovative consumers. Rather than launching a product on the market, it is important that innovation is well marketed. The first owner of a category gains more benefit from innovation. In addition, innovation is related to social impact rather than genetic codes (Hirschman, 1980: 283–285). Innovation is defined as cognitive and sensory innovation in two different ways. Cognitive innovation means the consumer's tendency to think rationally, problem-solving or other mental exercises. These mental activities encourage the consumer to be interested in new experiences. Cognitive innovators establish a relationship between goal and outcome, and ultimately assess how much they can meet their goals. Sensory innovation, is the consumer's feelings that are encouraging for new experiences. Sensory innovators do not hesitate to take more risks in choosing new things and do not make rational assessments to meet their needs (Dobre et al., 2009: 22–23). There are three different situations regarding the adoption of innovations. In the first case, the consumer is unaware of the gain from innovation if he does not yet understand what innovation promises. Innovation is capable of solving a major consumer problem, but the consumer is not aware of it. In the latter case, the consumer knows which problem of innovation will solve and has accepted the opinion of innovation, but has not purchased or used it. Therefore, he is aware of his earnings but he has not yet adopted innovation. In the third stage, buying and using behavior was realized. Innovation is adopted by the consumer (Hirschman, 1980: 293). For innovative consumers, the degree of acceptance of innovations is quite high. In other words, innovators form this group, which first adopts and uses new products. In simplified terms, innovative consumers accept new products and services rather early in their social environment compared to other consumers (Martens, 2010: 7). Innovators are known to be younger, more educated, from higher socioeconomic status, with money, and with a tendency to take risks when compared to other consumers. When compared with other social groups; innovators are in a hurry to try a new product that have the financial resources to meet the financial losses (Brancaleone & Gountas, 2007: 522). Innovative consumers serve as an important means of spreading new products or technologies. Individuals are not always motivated by a functional benefit when adopting an innovation. People can be motivated by more reputation or pretentiousness in their social environment, or they can be motivated to learn new things or just enjoy themselves. If a person is able to adopt innovations for pretension or social status, it is contradictory to give a definition that he is not influenced by others. Consumer innovation is the tendency of the consumer to adopt innovations (Tellis, Yin, & Bell, 2009: 1). With this definition, consumer innovation has moved to the point where it has the potential to do the action, not to do it.

2.2 *Opinion Leaders*

Opinion leaders are the other type of consumer among reference groups. One of the first definitions of opinion leaders concerns the tendency of other consumers to influence purchasing decisions. Opinion leaders, intentionally influence other consumers and their purchasing decisions. This is a feature that separates them from innovators (Martens, 2010: 8). The most important characteristic of the opinion leaders is that they have long-term interests, personal experiences and expertise about a particular product or service category. Consumers frequently gets advice from an opinion leader when making a purchase decision about any product or service category (Fitzmaurice, 2011: 71). The manufacturers of different products and brands have to determine how to reach and influence the opinion leaders in reference groups where group influence is strong. These people who have an impact on other people, use information about certain products and product groups, recommendations, opinions and communicate with other people communicate. With the increasing popularity of social networking sites, WOM messages are spreading very quickly, brand knowledge and awareness can be created by introducing products and services through opinion leaders. Opinion leaders play a guiding and influential role in consumers' quest for information in the purchasing process. Among the main characteristics of opinion leaders; levels of product engagement, wear, brand awareness, product information and individualization features income (Chan & Misra, 1990: 53). Opinion seekers show different characteristics than opinion leaders. Opinion leaders can be loyal brand users with higher brand awareness, where as opinion seekers, have high price sensitivity and do not focus on a particular product group. Although they follow different product groups, they are in search of information for their own needs. They search for opinions on social media, follow best-selling products and well-known brands. Many studies have been conducted to measure their characteristics (Bearden, Netemeyer, & Haws, 2011). The behavior of opinion leaders and opinion seekers was examined through online and off line communication and as a result of the research it was found that opinion seekers were much more active and more investigative in online communication environments. It has been observed that the role of opinion leaders in the dissemination and acceptance of technological innovations has been investigated (Cho, Junseok, & Lee, 2012: 97). Increasing the awareness of the brands by consumers and raising awareness will increase their success. In this way, brand loyalty can be created in a competitive environment as a priority to be preferred and an emotional connection can be established with the consumer. The sharing of opinion leaders and opinion seekers will play an important part in emotional bonding and establishing confidence among consumers. Because they are close concepts, opinion leadership and market mavens are sometimes confusing. Opinion leaders can influence people's attitudes, thoughts and behavior through personal communication (Hoffmann & Soyez, 2010). Market mavens are people with more general information about products and prices (Ruvio & Shoham, 2007: 705).

2.3 Market Mavens

Market maven studies was first carried out by Feick and Price. In the study, it was determined that market mavens exist and that other consumers are familiar with market mavens, but in terms of socioeconomic and demographic characteristics, market mavens do not differ significantly from other consumer groups (Feick & Price, 1987: 93). They follow new products and brands on the market. Even if they don't have detailed information about the products, they follow the campaigns, and they know where the products are best sold. They also try to disseminate this information through eWOM. They are the ones who have gained the trust of other consumers and have been tested for accuracy (Puspa & Rahardja, 2009: 25). Unlike opinion leaders and innovators, the personal effects of market mavens on consumers are in a wide range of products and brands (Abratt, Nel, & Nezer, 1995: 32). In this respect, market mavens' tendency to obtain information is consistent across the broad product range. They are stronger in perceiving price-quality relationships and have broader association skills with clarity, awareness, trial and acceptance (Williams & Slama, 1995: 5). These characteristics of market mavens make them attractive toward retailers and large industries as they can produce and sell durable and fast consumer goods like yogurt, drinks, laundry detergent, razor, etc. (Brancaleone & Gountas, 2007: 522). Although they have similar demographic characteristics as opinion leaders and innovators, are more likely to be female, with lower level of education than non-market mavens, who come from certain subcultural groups (Williams & Slama, 1995: 5). Market mavens are extroverted, perfectionist, have general market knowledge of the product, discuss with other consumers, and are willing to share information (Martens, 2010). It is necessary to characterize market mavens as not only consumers that provide useful information to other consumer, but also as smart shoppers (Slama, Nataraajan, & Williams, 1992). They are also observed to buy and try new products before other consumers. In this respect, they emphasize the need to see market mavens as a marketing opportunity (Goldsmith et al., 2003). Market mavens provide information to the market as a social role, so their interest in the market affects information sharing with other consumers and consumer perspectives on market mavens (Slama & Williams, 1990). In this context, market mavens collect information on various subjects in order to transfer them to other consumers about quality products, prices, discounts, product availability and quality of store employees (Brancaleone & Gountas, 2007). In different studies, it is suggested that personality traits are more functional than demographic traits in defining market mavens (Clark, Goldsmith, & Goldsmith, 2008: 245). The personal characteristics of market mavens have been tried to be determined by some researches. The five basic personality traits extroversion, neuroticism (emotional balance), openness to experience, compliance and self-discipline (responsibility) were evaluated with features (McCrea & Costa, 2003). The dimension of openness to experience puts people at a point on the line of being open to new experiences and strict adherence to habits. People who are constantly in search of different and varied experiences will receive high scores from the dimension of openness to experience. With WOM communication, the extraversion factor is intertwined. Therefore, market worms are expected to be more extroverted (Goodey & East, 2008: 269). Compared to other consumers, market wolves were found to have high scores of openness and extroversion factors from experience of five basic personality traits, while low scores of compatibility, self-discipline and emotional balance were found to be high (Mooradian, 1996). According to some other studies, other personality traits that define market worms are high self-esteem, tendency to be compatible, need to be indigenous and higher sensitivity to interpersonal relationships (Ruvio & Shoham, 2007: 706). Compliance was found to be easygoing, reliable, helpful, generous, accepting and good-natured. Individuals with high scores in compliance dimension are more likely to cooperate and individuals with low scores are more prone to competition. Market wolves are in a position to provide information about markets and help in this regard. Therefore, market worms are expected to have a positive relationship with the compatibility factor. The self-discipline dimension is related to how organized, controlled, disciplined, organized and success-oriented individuals are. Unstable people tend to change their products constantly. As a result, they are likely to have more information about the products to share. However, people with higher awareness may think that it is their social duty to transmit their knowledge about products to other people and share this information with others compared to indecisive people (Goodey & East, 2008: 269). Market mavens have more extensive with general market knowledge and more willing to disseminate information. Therefore, it is stated that market worms are more effective in sharing information within reference groups (Walsh & Elsner, 2012: 75). However, they do not have specific knowledge about products, such as opinion leaders. Like opinion leaders, they do not tend to be effective in a particular product category. Market mavens create trust and commitment to a certain object, business, brand, product, service, market place through the transfer of personal information with the effect on other people (Puspa & Rahardja, 2009: 26). If new information from consumers is positive, trust builds up and buying behavior occurs. Confidence persists and loyalty occurs according to satisfaction. The effect of market mavens here is whether it affects trust and commitment positively or negatively with the information content it shares. Considering this effect of market mavens, businesses, companies, marketing departments, etc. The importance of communication with market wolves emerges. On the other hand, market worms significantly reduce the communication costs of enterprises. Thus, market mavens gain value for businesses as a cheap advertising tool that shares information. These features also reveal the effect of WOM communication (Martens, 2010: 12–13). Market mavens have been studied in different studies in terms of their demographic, behavioral and psychological characteristics (Clark et al., 2008). In Poland, the market worm group was determined to be younger than non-market worm group, other demographic characteristics were measured on the basis of any difference (Chelminski & Coulter, 2002: 85). They found that the market mavens in Germany did not differ from the non-market mavens in terms of demographic characteristics (Clark & Goldsmith, 2005: 297). In Australia, it is found that those who are market-founded are more educated and have higher income than those who do not, but there is no difference in terms of gender (Brancaleone & Gountas, 2007: 526). In England, it has been found that female market worms are younger and less educated than men and the market worm scale tends to produce more meaningful results for female market worms (Goodey & East, 2008: 274).

3 E-WoM Marketing and Viral Advertising

WOM is a form of non-commercial verbal communication between two or more consumers interested in a brand, good or service (Arndt, 1967: 189). WOM is seen as an alternative to advertising or as a complement to traditional marketing communication techniques. Certain consumer groups play an active role in influencing the personal preferences of consumers (Clark & Goldsmith, 2005). WOM's strength in interpersonal interaction makes it a focal point for marketers. The strengthening of this interaction has brought innovation, opinion leaders and market wolves to the forefront (Chelminski & Coulter, 2007). Retailers strive to maximize their sales and profitability by attracting customers to their businesses. From this point of view, it is among the aims of spending less, exhibiting less marketing efforts and realizing positive WOM (Goldsmith et al., 2003). It plays an important part in shaping consumer attitudes and behaviors (Abratt et al., 1995). With WOM communication being replaced by communication on the internet, the importance of communication in the web has become more and more eWOM. They stated that the most valuable customers for businesses are those that create WOM effect to new customers rather than those who shop from that business (Kumar, Petersen, & Leone, 2007). eWOM, identifies all positive and negative comments made by existing or previous customers in forums and other social media platforms about a product or brand in online communities (Hennig-Thurau, Gwinner, Walsh, & Gremler, 2004). With the rapid increase in internet usage, shopping over the Internet has also increased rapidly, and eWOM has become widespread and accelerated marketing with this digital environment (Yang, 2013). The Internet has removed the boundaries and brought more active presence in chat rooms, referral sites and consumer evaluation corners according to the common interest of consumers all over the world. Developments in the digital environment have forced companies to develop more active marketing strategies for this environment. While firms have the opportunity to obtain very important information from consumer opinions and evaluations, they can also actively provide guidance in these communities (Mayzlin, 2006: 155). Consumers' opinions, decisions and buying behaviors that are declared outside the internet environment are based on the information they obtain from the internet environment (Jeong & Jang, 2011: 356). Therefore, it is stated that eWOM is a main source of information in determining the consumption behaviors of people. Together with eWOM, verbal and offline communication on product quality is replaced by online communication (Chevalier & Mayzlin, 2006: 345). The reason for the increasing importance of eWOM; consumers are now in more contact with each other. Consumers constantly tell each other about their experiences in daily life. In addition, it is now easier to receive and give advice via eWOM communication over the internet, allowing consumers to communicate not only with people they know, but also with foreign people and at the same time to access the information they want. Today, it is known that efforts to guide consumers by using various media channels have been maximized. In this environment, eWOM allows people to promote their products, brands and opinions worldwide, thus increasing their importance. Some brands even invite their customers to make their own advertisements, while others try to get support from their customers by listing their products and brands to support them. Organizations wishing to benefit from eWOM should carefully follow the emerging eWOM applications. This is due to the increasing use of eWOM and the introduction of new applications. In this respect, it is important that product representatives have sufficient knowledge of how eWOM works. They should also be able to evaluate the contribution of consumer communication to the company (Cruz & Fill, 2008). Today, social networking sites are often used for product and brand information and for improving brand value. It is even seen by marketers as a new and important viral advertising channel. As a result, brands using social media can directly connect with consumers and increase not only the ability to reach more customers, but also the likelihood of customers engaging in a buying activity (Nigam, 2012: 82). Market mavens also disseminates the information they have learned through eWOM (Puspa & Rahardja, 2009: 25). The messages presented on social media platforms both in the communication between consumers and the communication established by brands with consumers have a significant impact on brand value building. As a matter of fact, the communication established by brands with consumers on social media platforms is effective in creating brand value. In this respect, the necessity of involving customers such as market mavens should not be overlooked on marketing communications (Zailskaite-Jakste & Kuvykaite, 2013). There are a number of factors that contribute to the formation of eWOM, which has an impact on brand value. These are basically divided into eWOM seeking and motivation to provide eWOM. There are different views on the types of motivations. The use of eWOM and the development of consumer awareness in this regard are made faster because digital technologies facilitate access to eWOM resources. In addition to acquiring information on purchasing, eWOM also has an impact on purchasing behavior. If businesses encourage consumers to convey the problems they have with regard to the product or service; they can both find ways to solve problems and reclaim consumers. The recovered consumer can either repurchase or share this return with other people (Dhar & Nowlis, 1999). When the eWOM search motivation of the consumers was examined, it was determined that the consumers showed various eWOM motivation. The motivation to obtain information about the purchase is to reduce the risk perception of the consumer and the time spent on the purchase of a product. Consumers avoid purchasing activities if there is a negative trend in eWOM shares; they stated that they share with relatives. Another of eWOM search motivations is Social Orientation through Knowledge. Consumers have some problems choosing a particular product from alternatives and getting information about it. eWOM comments published in online communities can reduce problems because information provided by other consumers is more reliable than from companies or marketers. Another eWOM search motivation is Product Information. They stated that because of the role of the Internet in facilitating interpersonal communication and activities, consumers are common to the shopping experiences of other consumers and they have a sense of belonging with these groups by investigating the comments of other consumers in online communities. Another eWOM search motivation is the Economic Incentive. They state that many platforms of opinions reward customers for direct reading of their comments directly or indirectly because of the motivational nature of economic incentives (Hennig-Thurau & Walsh, 2003). According to another approach, product characteristics are getting more and more complex and product-related features are increasing even in the most ordinary products and the average consumer's ability to evaluate these product features is limited. Therefore, consumers rely more on eWOM as a product information source (Godes et al., 2005). Another motivation of consumers in eWOM research is Community Membership. Consumers often use the virtual environment to interact with other individuals and share their own consumption experiences. Community membership triggers consumers to join eWOM through factors such as feeling belonging to a community and being aware of new products entering the market (Granitz & Ward, 1996: 161). Being a member of a virtual community one can provide social benefits to other consumer with the elements of belonging and social integration; Therefore, the participation of consumers in online communities and feeling belonging shows that they are in eWOM communication (McWilliam, 2000: 43). eWOM marketing activities can be done in various ways. One of the most used is brand blogs. The blog is gün a blog-like website created by people who write their own things in their own way madden without requiring technical knowledge. Blogs can be preferred because they are cheap and free, do not require special software for installation, easily update and disseminate content, do not require computer knowledge at the level of expertise, add any comments, create dialogue between reader and writer (Goodfellow & Graham, 2007: 396). eWOM have positive impact on people and persuades them to buy products or services more than other ads. As they usually rely on what they hear directly from others (Maisam & Mahsa, 2016: 20). The messages transmitted by the source to the recipient during the eWOM process can be positive or negative. If the consumer is satisfied with the product or service he/she is using, a positive eWOM occurs, and if there is dissatisfaction, a negative eWOM occurs. Negative eWOM is the consumer reaction resulting from dissatisfaction (Richins, 1983: 69). It has a stronger effect than positive eWOM, and the negative word-of-mouth marketing continues even when the effect is disproved (Chioasa & Anastasiei, 2017: 159). Another concept in the sub-category of eWOM marketing is buzz marketing. Buzz marketing has become one of the most popular marketing trends in the globalizing world. It is generally referred to as whisper, gossip or rumor marketing. Thus, instead of spending millions in advertising campaigns, marketers began to take advantage of buzz marketing and began to save money (Notarantonio & Quigley, 2009: 456). Buzz marketing is the most important features of being interpersonal, having news value, the product, the opinion or brand is exciting, can be given to spread quickly. Reliability is at the top of the features that make eWOM powerful and reliable than advertising and other marketing methods (Oosterwijk &

Loeffen, 2005: 23–24). The ability of market mavens to become reliable emerges as an advantage for buzz marketing. In terms of market wolves, another source of information is viral marketing activities. The concept of viral marketing was first introduced in 1997 by Juvertson and Draper, focusing on the similarity in the spread of the electronic message and the spread of the virus. In viral marketing, it is aimed to spread the advertising messages among people rapidly and to create recognition of the brand over a wide audience (Woerndl, Papagiannidis, Bourlakis, & Li, 2008: 34). Viral marketing is similar to viruses due to its spreading properties. Viral marketing is the transmission of positive or negative information about the goods and services purchased and used by consumers to their friends, relatives and other familiar people in the internet environment. Since viral messages are carried out on the Internet, the effect is high. Viral marketing is beneficial for businesses. Market maven use effective viral marketing strategies to spread messages about businesses, goods and services to the people around them. Effective viral marketing strategies need to send goods and services sent to consumers free of charge, provide easy message to consumers, spread easily to large audiences, have something to convince consumers to make viral marketing possible. It is important to take advantage of, such as the use of different sources for viral marketing. In the study of Iacobucci and Hopkins (1992), it was revealed that the information transmitted virally differed between product categories. For example, they report that information on personal products is less rapid than information on non-personal products. In viral marketing, it performs the task of reducing the perceived risk in the purchasing process since information is obtained from the surrounding people and their experiences are utilized. The higher the threat in the purchasing process, the more the consumer wants to have information. Viral marketing is a risk reducing factor for consumers. The group of consumers, defined as market mavens with knowledge of the market, conducts viral marketing to other consumers. Most people who transmit information in viral marketing are those who founded the market. As the application area of viral marketing, market mavens have the knowledge about many goods and their shopping places, then they start the conversation on these topics and by answering the questions asked and provide market information to the consumers. In this type of marketing, the communication style used is informal as messages are spread by different channels such as email, chat rooms and discussion forums. From text and images to microsoft power point files, adobe flash animations, you tube video clips and so on, can cover various types of content (Xavier & Summer, 2009: 22). Social media, blogs, web sites, personal pages, etc. When the message is transmitted through channels, the number of people reached by passing the message from the consumer to the consumer increases. As information about goods and services is shared, the chances of marketing to potential consumers increase. The consumers who received marketing message stated that they were willing to forward the messages, if they came from their valuable people and their close circles and if they were supported by those people to send the messages to others (Yang, Liu, & Zhou, 2011). As marketers know more about market mavens, they will be more successful in implementing marketing strategies and influencing consumers (Clark et al., 2008). Retailers who want to publicize their goods and services with viral messages are trying to reach market mavens (Abrat et al., 1995). Market mavens are encouraged by businesses to share information with other consumers and spread their emotions through viral messages (Walsh, Gwinner, & Swanson, 2004). Market mavens visit businesses' websites to make various comments and contribute to other customers (Laughlin & MacDonald, 2010). Market mavens like to convey e-mails that other consumers think they will like and see it as a social responsibility (Phelps, Lewis, Mobilio, Perry, & Raman, 2004). Viral marketing and market mavens play an important role in retailing.

4 Consumer Based Brand Value

The concept of brand value was introduced by U.S. advertisers in the 1980s. The brand value is seen as the sum of the assets and liabilities associated with the brand's name and symbols, of the products and services offered by the brand determine the increase or decrease of this value. Brand value is the result of positive relationships creating value for the company and consumers. Companies with high brand value are associated with outstanding performance, includes sustainable price returns, inflexible price sensitivity, high market shares and successful expansion in new categories (Aaker, 1992). Strong brand value helps the company to establish itself in the market and identify itself. It also leads to higher profit margins and more corporate cooperation projects in measuring total brand value. The correct use of brand value through eWOM through social networking sites can add significant value to manufacturers, retailers and brand consumers (Nigam, 2012: 85). There are studies conducted on the classification of consumer based brand value dimensions. According to a study, four dimensions of brand value are brand loyalty, perceived quality, brand awareness/association and general brand value (Yoo & Donthu, 2001). According to another study, brand value has five dimensions: brand loyalty, brand awareness, perceived quality, brand association and registered brand assets (Aaker, 1992). In another study, the trust dimension was added to the brand and these dimensions were increased to six (Zahoor & Qureshi, 2017). The dimensions that make up brand value are generally handled within this framework. Brand loyalty, one of the dimensions that make up brand value, is a positive attitude for a brand in consumer preferences. Brand loyalty is the commitment of users to re-purchase or otherwise use the brand. Perceived quality is the perception of the customer about the expectations, alternatives and targets from the brand and the feelings toward the brand. In terms of eWOM brand communication, perceived quality refers to the ability of members to fulfill the expectation of a brand. Another concept that is important in terms of brand value is trust in the brand. The reliability of the brand is important in establishing long-term business relationships and in eliminating uncertainty in partnership structures and in situations that pose a risk to the consumer. When a continuous relationship is established between the consumer and the brand, trust increases along with brand awareness (Brogi et al., 2013: 4). Brand Image represents consumers' impressions, beliefs and everything they feel about a brand (del Bosque, Martin, & Collado, 2006: 412). The brand image affects the decision-making process of the consumer and this contributes to the increase of the brand value.

A strong and positive brand image enables the brand to be differentiated in the eyes of the consumers positively. Brand consciousness is the revival of a brand name in the mind of consumers by recognizing and remembering the brand (Zahoor & Qureshi, 2017: 51). Businesses are primarily trying to raise brand awareness in their marketing efforts. Since raising brand awareness alone will not be beneficial to the business, brands offer additional value that competitors do not offer to the consumer. Digital brand awareness is that online users recognize and remember the brand in the digital environment. Brand associations are the symbols, images, celebrities and so on. Explains emotions, beliefs and information that occur in the minds of the brand through the brand (Brogi et al., 2013: 4). Brand association is effective in creating brand loyalty by positively influencing purchasing decisions (Nigam, 2012: 82). Brand associations are sometimes made up of a wide range of specific features of the advertised product. If a classification of brand associations is found; brand qualifications, benefits and attitudes (Crayton, 2013: 27–28).

5 Conclusion

Nowadays, with the development of digital environment, the communication and marketing channels of businesses with consumers are changing. Consumers are increasingly confronted with a broad range of products and services in the digital environment, so market worms that emit eWOM messages in the digital environment play an increasingly active role in the interaction between businesses and consumers. Those who are described as market mavens are who work as a reference group, providing information to consumers around them about products and prices. These are individuals who have knowledge and share information with consumers about product types, shopping places and markets that respond to the wants and demands of consumers. Market mavens represent a particular group of consumers who are interested in every stage of the product life cycle, with high level of knowledge and active communication about both the product and the market, even if they do not buy or use a product. As a rising value, market mavens emerge as an important group that affects consumers through eWOM and through spreading viral advertisements. They sometimes share information without being asked by consumers. Brands that are aware of the eWOM effect and the importance of viral marketing should ensure that they produce content and communicate information to consumers using consumer reference groups, especially market mavens, rather than advertising. It is defined as eWOM communication for consumers, convey their positive and negative experiences and information about their products and services to other consumers on internet. With eWOM, viral marketing serves as an important tool in creating brand value (Nigam, 2012: 87). Consumer reference groups are evaluated in a variety of ways. Reference groups are considered in three groups as opinion leaders, innovators and market mavens (Ruvivo & Shoham, 2007). Although the experiences and behaviors of these three groups vary, they tend to share their knowledge and experiences with consumers. Opinion leaders have a wide range of opinions about specific product categories and make recommendations on them; innovators are particularly experienced in new products (Goldsmith et al.,

2003). In contrast, market mavens have more general information about the market which they share with consumers (Fitzmaurice, 2011).

Viral communication studies with eWOM are perceived to be similar and are often used interchangeably. In this respect, viral marketing plays an active part in every stage of the consumer's purchasing decision process. Particularly in marketing communication, market mavens use viral message channels effectively. In the purchasing decision process of consumers, viral messages on virtual communication networks have a positive impact on the buying behavior of consumers as it is a source of information for eWOM. From business marketing point of view, the importance given to eWOM communication is less costly and relatively more efficient. Consumers are seeking information in the purchasing decision process. At this stage, market mavens play an important role in informing consumers and guiding them in purchasing decisions. Market mavens are facilitating factors via eWOM such as reducing the perceived risk for consumers, shortening the process of obtaining information, learning how to use the product, learning which products are new in the markets, reducing the negative effects after purchasing and belonging to a virtual community. Also, market mavens can influence consumers via eWOM channels such as on social media, email, consumer chat rooms, discussion forms and newsgroups. Through social media, consumers are able to obtain a wide range of information about the products and services they are interested in, from their close friends and acquaintances. For brands, various recommendations can be made to exploit the eWOM effect and viral communication effects of market mavens. Brands should use social media consciously and actively, investigate market wolves, contact them and conduct communication activities through them. For this, firstly, accounts should be opened in the social network environment, the consumers of the brand should be reached through the social network and consumers should be encouraged to follow the social media accounts of the brand. Candidates from social media followers who may be market mavens should be identified, focused on these people and encouraged to be a representative of the brand. They should encourage market mavens to talk about the brand's goods and services by providing them with special privileges, gifts and discounts. Rather than making expensive advertising on mass media, it can be more economical and more effective to promote it through eWOM communication and viral advertising. In addition, the brand's content generation and eWOM effect will create a platform that facilitates the perception of positive brand building, and the usefulness of these contents for consumers will increase the profitability of the brand and save time. eWOM communication, which is made by highly satisfied consumers, is especially important for businesses in the field of service and retailing to find new customers and retain their existing customers.

In future studies, market mavens and non-market mavens can be separated and examined for the effects of opinion leaders and innovators, as well as the impact of eWOM and viral marketing on different reference groups in the community.

References

Aaker, D. A. (1992). Managing brand equity: Capitalizing on the value of a brand name. *Journal of Marketing, 56*(2), 125–128.

Abratt, R., Nel, D., & Nezer, C. (1995). Role of the Market maven in retailing: A general marketplace influencer. *Journal of Business and Psychology, 10*(1), 31–55.

Arndt, J. (1967). The role of product-related conversations in the diffusion of a new product. *Journal of Marketing Research, 4*(3), 291–295.

Assink, M. (2006). Inhibitors of disruptive innovation capability: A conceptual model. *European Journal of Innovation Management, 9*(2), 215–233.

Bearden, O. W., Netemeyer, R. G., & Haws, K. L. (2011). *Handbook of marketing scale, multi item measures for marketing and consumer behaviour research* (3rd ed., pp. 106–107). Thousand Oaks, CA: Sage.

Bearden, O. W., Netemeyer, R. G., & Teel, J. E. (1989). Measurement of consumer susceptibility to interpersonal influence. *Journal of Consumer Research, 15*(4), 473–481.

Brancaleone, V., & Gountas, J. (2007). Personality characteristics of Market mavens. *Advances in Consumer Research, 34,* 522–527.

Brogi, S., Calabrese, A., Campisi, D., Capece, G., Costa, R., & Di Pillo, F. (2013). The effects of online brand communities on brand equity in the luxury fashion industry. *International Journal of Engineering Business Management Special Issue on Innovations in Fashion Industry, 5*(32), 1–9.

Buttle, F. A. (1998). Word of mouth: Understanding and managing referral marketing. *Journal of Strategic Marketing, 6,* 241–254.

Chan, K. K., & Misra, S. (1990). Characteristics of the opinion leader: A new dimension. *Journal of Advertising, 19*(3), 53–60. Published online: 29 May 2013.

Chelminski, P., & Coulter, R. (2002). Examining Polish market mavens and their attitudes toward advertising. *Journal of East-West Business, 8*(1), 77–90.

Chelminski, P., & Coulter, A. R. (2007). On market mavens and consumer self-confidence: A cross-cultural study. *Psychology & Marketing, 24*(1), 69–91.

Chen, J. (2002). *Chinese identification, acculturation and conspicuous consumption in a multicultural society: A theoretical framework and empirical evidence* (Phd. Dissertation), The University of Guelph, UMI.

Chevalier, J. A., & Mayzlin, D. (2006). The effect of word of mouth on sales: Online book reviews. *Journal of Marketing Research, 43*(3), 345–354.

Chiosa, A. R., & Anastasiei, B. (2017). Negative word-of-mouth: Exploring the impact of adverse messages on consumers' reactions on facebook. *Review of Economic Business Studies, 10*(2), 157–173.

Cho, Y., Junseok, H., & Lee, D. (2012, January). Identification of effective opinion leaders in the diffusion of technological innovation: A social network approach. *Technological Forecasting and Social Change, 79*(1), 97–106.

Clark, R. A., & Goldsmith, R. E. (2005). Market mavens: Psychological influences. *Psychology & Marketing, 22*(4), 289–312.

Clark, R. A., Goldsmith, R. E., & Goldsmith, E. B. (2008, May–June). Market mavenism and consumer self-confidence. *Journal of Consumer Behavior, 7*, 239–248.

Crayton, M. K. (2013). *Examining the relationship between the usefulness of multimedia messaging services and brand equity: A conceptual framework* (Doctor of Philosophy), Capella University. Minneapolis.

Cruz, D., & Fill, C. (2008). Evaluating viral marketing: İsolating the key criteria. *Marketing Intelligence & Planning, 26*(7), 743–758.

Das, S. (2020). Innovations in digital banking service brand equity and millennial consumerism. *Digital transformation and innovative services for business and learning* (pp. 62–79). IGI Global. https://doi.org/10.4018/978-1-7998-5175-2.ch004.

Das, S. (2021). *Search engine optimization and marketing: A recipe for success in digital marketing* (1st ed.). CRC Press. https://doi.org/10.1201/978042 9298509.

Das, S., & Mondal, S. (2016). *Yours digitally: Patanjali.* Available at SSRN 2737299.

Das, S., Mondal, S. R., Sahoo, K. K, Nayyar, A., & Musunuru, K. (2018). Study on impact of socioeconomic make up of Facebook users on purchasing behavior. *Revista Espacios, 39*, 28–42. Retrieved from http://www.revistaes pacios.com/a18v39n33/18393328.html.

Das, S., & Nayyar, A. (2019). Innovative ideas to manage urban traffic congestion in cognitive cities. In *Driving the development, management, and sustainability of cognitive cities* (pp. 139–162). IGI Global. https://doi.org/ 10.4018/978-1-5225-8085-0.ch006.

Das, S., & Nayyar, A. (2020). Effect of consumer green behavior perspective on green unwavering across various retail configurations. In *Green marketing*

as a positive driver toward business sustainability (pp. 96–124). IGI Global. https://doi.org/10.4018/978-1-5225-9558-8.ch005.
Das, S., Nayyar, A., & Singh, I. (2019). An assessment of forerunners for customer loyalty in the selected financial sector by SEM approach toward their effect on business. *Data Technologies and Applications, 53*(4), 546–561. https://doi.org/10.1108/DTA-04-2019-0059.
del Bosque, I. A. R., San Martín, H., & Collado, J. (2006). The role of expectations in the consumer satisfaction formation process: Empirical evidence in the travel agency sector. *Tourism Management, 27*(3), 410–419.
Dhar, R., & Nowlis, S. M. (1999). The effect of time pressure on consumer choice deferral. *Journal of Consumer Research, 25*(4), 369–384.
Dobre, C., Dragomir, A., & Preda, G. (2009). Consumer innovativeness: A marketing approach. *Management & Marketing, 4*(2), 19–34.
Duy, N. T., Mondal, S. R., Van, N. T. T., Dzung, P. T., Minh, D. X. H., & Das, S. (2020). A study on the role of web 4.0 and 5.0 in the sustainable tourism ecosystem of Ho Chi Minh City, Vietnam. *Sustainability, 12*(17), 7140. https://doi.org/10.3390/su12177140.
Escalas, J. E., & Bettman, J. R. (2005). Self-construal, reference groups and brand meaning. *Journal of Consumer Research, 32*(3), 378–389.
Feick, L. F., & Price, L. L. (1987, January). The market maven: A diffuser of marketplace information. *Journal of Marketing, 51,* 83–97.
Fitzmaurice, J. (2011). Market mavens' motivations to acquire information. *The Marketing Management Journal, 21*(1), 71–83.
Godes, D., Dina, M., Yubo, C., Sanjiv, D., Chrysanthos, D., & Bruce, P. (2005). The firm's management of social interactions. *Marketing Letters, 16*(3/4), 415–428.
Goldsmith, R. E., Flynn, L. R., & Goldsmith, E. B. (2003). Innovative consumers and market mavens. *Journal of Marketing Theory and Practice, 11*(4), 54–64.
Goodey, C., & East, R. (2008). Testing the market maven concept. *Journal of Marketing Management, 24*(3–4), 265–282.
Goodfellow, T., & Graham, S. (2007). The blog as a high impact institutional communication tool. *The Electronic Library, 25*(4), 395–400.
Granitz, N. A., & Ward, J. C. (1996). Virtual community: A sociocognitive analysis. *Advances in Consumer Research, 23*(2), 161–166.
Gupta, D. K., Jena, D., Samantaray, A. K., & Das, S. (2019). HRD climate in selected public sector banks in India. *Revista ESPACIOS, 40*(11), 14.
Hamilton, R. W. (2003). What do people suggest what they don't want? Using context effects to influence other's choices. *Journal Consumer Research, 29*(4), 492–506.
Hennig-Thurau, T., Gwinner, K. P., Walsh, G., & Gremler, D. D. (2004). Electronic word-of-mouth via consumer-opinion platforms: What motivates

consumers to articulate themselves on the internet? *Journal of Interactive Marketing, 18*(1), 38–52.

Hennig-Thurau, T., & Walsh, G. (2003). Electronic word-of-mouth: Motives for and consequences of reading customer articulations on the internet. *International Journal of Electronic Commerce, 8*(2), 51–74.

Hirschman, E. C., (1980, December). Innovativeness, novelty seeking, and consumer creativity. *Journal Of Consumer Research, 7,* 283–295.

Hoffmann, S., & Soyez, K. (2010). A cognitive model to predict domain-specific consumer innovativeness. *Journal of Business Research, 63,* 778–785.

Iacobucci, D., & Hopkins, N. (1992). Modeling dyadic interactions and networks in marketing. *Journal of Marketing Research, 29*(1), 5–17.

Jain, S., Jain, V., & Das, S. (2018). Relationship analysis between emotional intelligence and service quality with special evidences from Indian banking sector. *Revista ESPACIOS, 39*(33), 3.

Jeong, E., & Jang, S. (2011). Restaurant experiences triggering positive electronic word-of-mouth (Ewom) motivations. *International Journal of Hospitality Management, 30,* 356–366.

Kotler, P. (2002). *Marketing management* (Millenium ed.). Upper Saddle River, NJ, USA: Pearson Prentice Hall.

Kumar, V., Petersen, A., & Leone, R. P. (2007). How valuable is word of mouth? *Harvard Business Review, 85*(10), 139–146.

Laughlin, J. D., & MacDonald, J. B. (2010). Identifying market mavens online by their social behaviors in community-generated media. *Academy of Marketing Studies Journal, 14*(1), 55–70.

Maisam, S., & Mahsa, R. (2016). Positive word-of–mouth marketing: Explaining the role of value congruity and brand love. *Journal of Competitiveness, 8*(1), 19–27.

Martens, M. (2010). *The effect of Market Mavens on trial probability: Does marketing communication affect market mavens* (Bachelor thesis marketing), Supervisor: Ms. N. Raassens, pp. 1–27.

Mayzlin, D. (2006). Promotional chat on the internet. *Marketing Science, 25*(2), 155–163.

McCrae, R. R., & Costa, P. T. (2003). *Personality in adulthood.* New York: Guilford Press.

McWilliam, G. (2000). Building stronger brands through online communities. *Sloan Management Review, 41*(1), 43–54.

Mohanty, P. C., Dash, M., Dash, M., & Das, S. (2019). A study on factors influencing training effectiveness. *Revista Espacios,* 40, 7–15. Retrieved from http://www.revistaespacios.com/a19v40n02/19400207.html.

Mondal, S. R. (2020). A systematic study for digital innovation in management education: An integrated approach towards problem-based learning in Vietnam. In *Digital innovations for customer engagement, management, and organizational improvement* (pp. 104–120). IGI Global. https://doi.org/10.4018/978-1-7998-5171-4.ch006.

Mondal, S., Das, S., Musunuru, K., & Dash, M. (2017). Study on the factors affecting customer purchase activity in retail stores by confirmatory factor analysis. *Revista Espacios, 38*(61), 30.

Mondal, S., Mall, M., Mishra, U. S., & Sahoo, K. (2017). Investigating the factors affecting customer purchase activity in retail stores. *Revista Espacios, 38*(57), 22.

Mondal, S., & Sahoo, K. K. (2020). A study of green building prospects on sustainable management decision making. In *Green building management and smart automation* (pp. 220–234). IGI Global. https://doi.org/10.4018/978-1-5225-9754-4.ch011.

Mooradian, T. A. (1996). The five factor model and market mavenism. *Advances in Consumer Research, 23,* 260–263.

Nadanyiova, M., & Das, S. (2020). Millennials as a target segment of socially responsible communication within the business strategy. *Littera Scripta, 13*(1), 119–134. https://doi.org/10.36708/Littera_Scripta2020/1/8.

Nigam, A. (2012). Influence of viral marketing on brand equity building with reference to online social networking sites networking sites. *International Journal of Computer Science & Management Studies, 12*(1), 82–88.

Notarantonio, E., & Quigley, C. (2009). The effectiveness of a buzz marketing approach compared to traditional advertising: An exploration. *Journal of Promotion Management, 15,* 455–464.

Oosterwijk, L., & Loeffen, A. (2005). *How to use buzz marketing effectively? A new marketing phenomenon explained and made practical* (Unpublished master thesis), University of Mälardalen, Västerås.

Phelps, J. E., Lewis, R., Mobilio, L., Perry, D., & Raman, N. (2004). Viral marketing or electronic word-of-mouth advertising: Examining consumer responses and motivations to pass along email. *Journal of Advertising Research, 44*(4), 333–348.

Puspa, J., & Rahardja, C. T. (2009). Market mavenism its contribution in building consumer's trust. *Gadjah Mada International Journal of Business, 11*(1), 21–35.

Richins, M. L. (1983). Negative word of mouth by dissatisfied consumers: A pilot study. *Journal of Marketing, 47,* 68–78.

Robertson, T. S. (1967). The process of innovation and the diffusion of innovation. *Journal of Marketing, 31*(1), 14–19.

Rogers, E. (1976). New product adoption and diffusion. *Journal of Consumer Research, 2*(4), 290–301.

Ruvio, A., & Shoham, A. (2007). Innovatiness, exploratory behavior, market mavenship, and opinion leadership: An empirical examination in the Asian context. *Pschology & Marketing, 24*(8), 703–722.

Sharma, E., & Das, S. (2020). Measuring impact of Indian ports on environment and effectiveness of remedial measures towards environmental pollution. *International Journal of Environment and Waste Management, 25*(3), 356–380. https://doi.org/10.1504/IJEWM.2019.10021787.

Singh, I., Nayyar, A., & Das, S. (2019). A study of antecedents of customer loyalty in banking & insurance sector and their impact on business performance. *Revista ESPACIOS, 40*(06), 11.

Singh, I., Nayyar, A., Le, D. H., & Das, S. (2019). A conceptual analysis of internet banking users' perceptions. An Indian perceptive. *Revista ESPACIOS, 40*(14), 1.

Singh, L. B., Mondal, S. R., & Das, S. (2020). Human resource practices & their observed significance for Indian SMEs. *Revista ESPACIOS, 41*(07). Retrieved from http://www.revistaespacios.com/a20v41n07/20410715.html.

Singh, S., & Das, S. (2018). Impact of post-merger and acquisition activities on the financial performance of banks: A study of Indian private sector and public sector banks. *Revista Espacios Magazine, 39*(26), 25.

Singh, S., Mondal, S., Singh, L. B., Sahoo, K. K., & Das, S. (2020). An empirical evidence study of consumer perception and socioeconomic profiles for digital stores in Vietnam. *Sustainability, 12*(5), 1716.

Slama, M. E., Nataraajan, R., & Williams, T. G. (1992). Market mavens and the relationship between smartbuying and information provision: An exploratory study. In V. L. Crittenden (ed.), *Developments in marketing science* (vol. 15, pp. 90–93). Cham: Springer.

Slama, M. E., & Williams, T. G. (1990). Generalization of market maven's information provision tendency across product categories. *Advances in Consumer Research, 17,* 48–52.

Slaughter, E. S. (1998). Models of construction innovation. *Journal of Construction Engineering and Management, 124,* 226–212.

Solomon, R. M. (2007). *Consumer behavior, buying, having, and being* (7th ed.). Upper Saddle River, NJ: Pearson-Prentice Hall.

Tellis, G., Yin, E., & Bell, S. (2009). Global consumer innovativeness: Cross-country differences and demographic commonalities. *Journal of International Marketing, 17* (2), 1–22.

Van, N. T. T., Vrana, V., Duy, N. T., Minh, D. X. H., Dzung, P. T., Mondal, S. R., & Das, S. (2020). The role of human–machine interactive devices for post-COVID-19 innovative sustainable tourism in Ho Chi Minh City, Vietnam. *Sustainability, 12*(22), 9523. https://doi.org/10.3390/su12229523.

Walsh, G., & Elsner, R. (2012). Improving referral management by quantifying market mavens' word of mouth value. *European Management Journal, 30,* 74–81.

Walsh, G., Gwinner, K. P., & Swanson, S. R. (2004). What makes Mavens Tick? Exploring motives of market mavens' initiation of information diffusion. *Journal of Consumer Marketing, 21*(2), 109–122.

Williams, T. G., & Slama, M. E. (1995). Market mavens' purchase decision evaluative criteria: Implications for brand and store promotion efforts. *Journal of Consumer Marketing, 12*(3), 4–21.

Woerndl, M., Papagiannidis, S., Bourlakis, M., & Li, F. (2008). Internet-induces marketing techniques: Critical factors in viral marketing campaigns. *Journal of Business Science and Applied Management, 3*(1), 34–45.

Xavier, L., & Summer, G. (2009). *Viral marketing communication: The internet word-of-mouth* (Unpublished master thesis), Blekinge Institute of Technology, Karlskrona.

Yang, H. (2013). Market mavens in social media: Examining young Chinese consumers' viral marketing attitude, eWom motive, and behavior. *Journal of Asia-Pacific Business, 14,* 154–178.

Yang, H. C., Liu, H., & Zhou, L. (2011). Predicting young chinese consumers mobile viral attitudes, intents and behavior, *Asia Pacific Journal of Marketing and Logistics, 24,* 59–77.

Yoo, B., & Donthu, N. (2001). Developing and validating a multidimensional consumer-based brand equity scale. *Journal of Business Research, 52,* 1–14.

Zahoor, S. Z., & Qureshi, I. H. (2017). Social media marketing and brand equity: A literature review. *IUP Journal of Marketing Management, 16*(1), 47–64.

Zailskaite-Jakste, L., & Kuvykaite, R. (2013). Communication in social media for brand equity building. *Economics and Management, 18*(1), 142–153.

CHAPTER 9

Integrated Marketing Communication: A Systematic Study from Media Channels to Digital Connectivity of New Age Consumers

Teena Mishra

1 Introduction

Since the creation of the World Wide Web by the Sir Tim Berners Lee around three billion people in the world are using online services for search and purchasing products, entertainment, and friends (Dave Chaffey and Fiona Ellis Chandwick). The key to get success of all digital marketing is well informed and acknowledged customers (Dodson, 2016). With this modern digital technology customers are able to share their experiences about the products and services which give feedback to the marketers. Customers have a freedom to share their reason of like or dislike particular products. Now, it is easy to enter in the market through internet. It is not tough time for the small business to reach customers. Internet marketing communication channel is the easiest and

T. Mishra (✉)
PSS Central Vocational Institute, Bhopal, India

S. Das and S. Gochhait (eds.), *Digital Entertainment*,
https://doi.org/10.1007/978-981-15-9724-4_9

fastest communication channel. Let's first discuss about the meaning of marketing communication channel.

2 Meaning of Marketing Communication Channel

To move any place we need vehicle. Similarly, to deliver message from customer to business or business to business or customer to customer is called marketing communication channel (Singh, Mondal, Singh, Sahoo, & Das, 2020). It is the component of the marketing campaign. 'Media channels can work and can fail' statement clarified that it is not necessary that every time business get success through the media channels. Success depends on how to integrate and deliver the message (Duy et al., 2020; Nadanyiova & Das, 2020; Mulhern, 2009; Van et al., 2020).

3 Meaning of Digital Marketing

Digital marketing connects marketers to customers with the use of online technology. Digital marketing means the use of the Internet, social media, search engines, and mobile devices and various other channels to attract, persuade, and make available products and services to the existing as well as new customers (Singh, Mondal, & Das, 2020).

Digital marketers who are using strategies and tactics have a well-known idea of how to achieve goals with the aid of digital marketing (Sharma & Das, 2020). Digital marketers are providing information links about online certification courses, educational courses, and various other products and services (Das, 2020). The trend of online educational services increased after the outbreak of COVID 19 2020. Digital marketing focused to a specific segment of the customers and it is interactive. It is rising day by day and includes various email ads, tweets and anything that is giving opportunity to take feedbacks of the customers (Das, 2021; Mondal, 2020).

4 Difference Between Internet Marketing and Digital Marketing

Internet marketing is providing adverting services on the internet whereas, digital marketing can do through mobile phones, smartphone app, or in a video game (Patlin, 2019), for example, banks are asking their customers to download bank app to take its various services.

5 Conventional Vs Digital Media Channels

Conventional media channels are various tools of the traditional media channels which are paid advertising, direct marketing, word of mouth, events, public relations, partnership, and joint ventures (Mondal & Sahoo, 2020). *Various digital media* tools are inbound marketing, content marketing, email marketing, search engine optimization, local search, social media, paid advertising, and mobile marketing and affiliate marketing (Talafuse, 2014). Now the question is which media channel should be used to brand the business? The answer is business can take advantage from using a mix of conventional and digital media (Das & Nayyar, 2020). It is not the issue which media individual or business persons are using, the purpose behind it matters a lot (Das, Nayyar, & Singh, 2019). There are various options available as traditional media, digital media, and new media. There are benefits of all types of media channels but the thing is what the kind of business is and what are their goals? Selection of the channel depends on the objective of using it (Das & Nayyar, 2019). The best way is to plan the strategies for using channels and to solve the purpose of message delivery. The ultimate success is on the hands of the target customers (Singh, Nayyar, Le, & Das, 2019). So, it's the plan and strategies of using media channels to attract and persuade customers. The actual difference between the traditional and digital media is where the message is going to appear (Mohanty, Dash, Dash, & Das, 2019). However, the message is same, but the way of presenting it is different. Various new media are nowadays using digital banners, google ads, social media, and blogs. When the business are using combined use of conventional as well as digital channels they are called as multi-channel marketing (Singh, Nayyar, & Das, 2019). Table 1 shows various marketing conventional and digital marketing channels.

6 Use of Digital Marketing Channels

In the recent trends companies are using digital marketing channels to achieve various goals. Some of the goals are mentioned below.

Table 1 Conventional and digital media channels

Conventional marketing communication channels	*Digital*
Print publications	Website
Radio	Blogs
Billboards	Podcasts
Signage	Social Media
Telephones	Email
Postal Service	Video
Events	Mobile
	Webinars

1. **To create brand awareness**: To create brand awareness digital marketing channels are the best technique. The reason is that it gives greater product name recall and creates familiarity with the business (Gupta, Jena, Samantaray, & Das, 2019). There are some specifications with the digital media channels. Various social media channels are Facebook, twitter, google, etc., promotes products and service easily with low cost. Channels like Instagram and Pinterest are easily promoting visual elements and the channels that promote videos are Youtube and Vimeo (Singh & Das, 2018).
2. **Generates sales**: Various business market their products and service to create sales. E-commerce becomes the part of everyone's life. It gives exposure to many people at one time. It is easy method of purchasing thus generate sales (Jain, Jain, & Das, 2018). Various online channels are; Amazon, Flip cart, Jabong, Snapdeal, Google display network, and Facebook, etc.
3. **Generates lead**: Digital Marketing used to generate leads. It attracts prospective customers and generates leads. Communication is direct and easy to do follow-up. Various ways are search engine optimization (SEO), Pay-Per-Click Advertising, Display Advertising, google search network, Blogging, Email-marketing, and others (Das, Mondal, Sahoo, Nayyar, & Musunuru, 2018).
4. **Educate customers**: Business inform and educate new as well as existing customers. Digital marketers convey the message to customers about benefits of products and services. They also help to resolve their problems. Blogging, content marketing, and press releases are useful for this reason (Mondal, Das, Musunuru, & Dash, 2017).

5. **Generate traffic**: Various websites are publishers of online ads, content writing, display advertising, for generating revenue. Digital marketing is best for content marketing and search engine optimization (Mondal, Mall, Mishra, & Sahoo, 2017).

6.1 Marketing Communication Strategies

Strategies give direction to accomplish a goal. In the marketing communication there are various tools to deliver message to the target audience. The best strategy is to use the appropriate channel through which message deliver successfully and effectively (Das & Mondal, 2016).

7 Principles of the Successful Marketing Strategy

There are three I principles of successful marketing strategy: Initiate, Iterate, and integrate (Dodson, 2016).

- *First 'I' Principle is Initiate*: Organizations have to be ready to take feedback from customers. Customers have all the information's but the thing is how an organization reveals information's like resources, channels, budget, and strategy.
- *Second 'I' Principle is Iterate*: Second principle iterates which represents alteration in response to the user interaction.
- *Third 'I' Principle is Integrate*: The third I principle is integrate, which means the efforts for integrating customers is the strategy of the business organization across digital channels.

8 Digital Marketing Channels

Michaelle Aguila (2019) suggested digital marketing channels that are effectively working. Each digital channel is exclusive in their features, objectives, and structure. Now the question is which channel gives the finest return on investment. Effective digital marketing channels are websites, email, social media, organic search, paid search, mobile, display ads. Lucy Alexander suggested about in what way digital marketing happens. Internet users are increasing and marketing focusing to connect with audience at the right place and at right time.

Marketing efforts that uses an electronic device or internet is known as digital marketing. Various digital channels are search engines, social media, websites, emails, etc., to connect with customers. Digital marketing is important for the business and business awareness. There are many options in the digital marketing that can use tools for example dashboards to monitor, analytics for the success and ROI of campaigns. There are various marketers who uses channels to convey message to the customers. Due to various numbers of channels available, it is vital to select best channel. The best way is to plan the campaigns in advance and discuss the various key points.

There are various digital marketing channels. To identify the right digital marketing tools it is necessary to understand the various marketing channels. Various types of marketing channels are organic, email-traffic, paid-search traffic, content marketing traffic, social media traffic, affiliate traffic, referral strategic, direct traffic, display advertising, and online public relations (Ryan, 2014) explained below.

1. *Organic*: It includes all websites traffic that are available through search engines for example, Google, Bing, Yahoo, etc. Whenever people search any information about product or topic then they discover the websites. For more traffic, it is better to invest in search engine optimization (SEO). All people are familiar with the internet. They used search engine to search out information's about the products and services.
2. *Email Traffic*: Email traffic includes the traffic during email messages. These email messages can be promotional e-mail campaigns, triggered email campaigns, and transactional email campaigns. Email-Marketing information's are collected from the potential customers to maintain the business for example, phone numbers and email address. Potential customers are getting updates and other information through the email. Email is effective means of digital marketing channels. Various in-house businesses are using email as marketing channel.
3. *Paid Search Traffic*: This creates traffic during paid ads on the search engines. It is also called to as search engine marketing (SEM) or pay-per-click (PPC). Pay-Per-Click (PPC) advertising is a paid advertising. In this advertising budget is within the control of business. It makes how to set budget to spend for one day, a week, or a month.

4. *Content Marketing Traffic*: This includes the website traffic from the blog, digital downloads, videos. It is also called as inbound marketing. Content marketing is one of the vigorous marketing campaigns in the digital field. It remains same during various changes in the marketplace. In the B2B business are increasing the content marketing. It includes graphics, videos, text, and others.
5. *Social Media Traffic*: Social media is the most popular media in the present scene. It includes traffic that receive form the social media networks, for example, facebook, twitter, instagram, snapchat, etc., traffic comes may be free or paid. Social Media Marketing (SMM). Facebook, Twitter, Linkedin, Google, are the social media sites. Digital marketers create brand awareness for their products and services through the social media marketing. The numbers of users are increasing day by day Shamsudeen Ibrahim and Ganeshbabu (2018).

   - *Facebook*: Facebook helps to reach any type of general consumer
   - *Twitter*: Twitter helps to generate speedily brand awareness.
   - *LinkedIn*: LinkedIn is for professionals, business owners, and other B2B *Google*: It is for new business as well as B2B and general social media user.
   - *Pinterest*: It is the social network which promotes brands and are having mostly female users. So if the target market is female dominated, then this will be the best option.

6. *Affiliates traffic:* Affiliate digital marketing channels receive traffic from the affiliate partners in exchange for a commission and popular networks for it are CJ.com, ShareASale.com, Rakuten,LinkShare. For wider reach and global activity there is no need to hire persons at different locations. Affiliate marketers can work on the basis of commissions. For example, pay-per-performance is the online business model.
7. *Referrals Traffic:*It includes all the traffic that receives from everything on the internet. It also includes apps. Various websites that are running sponsored blog post or digital newspaper, display ads are included in the referrals traffic.
8. *Direct Traffic:* This includes all websites that directly comes to the other websites. Direct traffic helps to compute some of the offline marketing efforts, for example, TV, radio, or events. For

example, on tv advertisement people watch about the website and they access the website by using the smart phone. The most important thing to create a digital marketing strategy is the objectives and performance metrics. It is necessary to measure and monitor the performance.

9. *Display Advertising:* Blogs, forums, and other websites are useful for various people. Display advertising helps to reach potential customers by placing advertisement on the third party sites. It includes banners, boxes, interactive ads, video, and other ads. It generates highly targeted traffic that converts into sales of products and services. Cost is based on CPI (Cost per Impression). It is cost effective than PPC.
10. *Online Public Relations:* Online public relations are a part of the content marketing and generally create promotional content. It is used to generate branding and creating traffic through Public relation network. Some channels require joining subscription and some are freely available.

9 Prepare and Allocate Budget for Digital Marketing

Cost and features of all the digital marketing channels are unique and its selections reveal the best marketing strategy of Lilach Bullock (2019). Thinking alone about print, electronic and outdoor marketing without using digital marketing shows out of the box in the present market. Marketing budget is the first thing that marketer jump to select the marketing communication channel. Selection of channel is not dependent on your interest only but also on your budget. Online marketing made easy and reasonable budget for many marketers. To make business visibility to the target audience multi-channel marketing strategy is the best strategy. Inbound marketing is to build awareness and convert customers into paying customers. Small business prefer inbound marketing that includes search, social, and content marketing. PPC adds click by the various people on Facebook, Google, or other social media in the search result. Poorly designed campaign can cost lot of money. Youtube is the online platform to educate audience, providing product information, sharing knowledge, and to engage audience. Videos are easy mode of communication for the audience. They can watch it any time and at any place. It can build a quality channel to connect people as an expert in the field. Email Marketing is to make this strategy effective where it is important to manage and divide the list and create a good attractive email ID. Event Marketing: Webinars and virtual events are possible at the cost of web conferencing for all businesses. The recent example of various virtual events and webinars occurred due to COVID 19. Develop a referral strategy that has customer's reviews on Google and others. Direct marketing through postcards, letter, email, telephone make the message reaches to the customer. Successful campaigns use strategies of multiple channels so that message reaches the customer. However, all channels do not carry same weights.

Selection depends on the goal that marketing personnel want to achieve. Thus, business needs to understand following things to develop quality marketing communication plans which are:

- Goals of campaign (awareness, sale)
- Target audience
- Value of the product or services
- Messages want to communicate
- Select best communication media.

If the marketer uses inbound technique like SEO, social media, and content creation for the website then there is no need to invest much budget on it. Money investment is less in these types of digital marketing. The real investment in these types of digital types of marketing is time investment. Outbound methods like online advertising have some expenses. Its cost depends on visibility that marketers want to see. Identify the budget. Budget allocation includes fees for digital manager, advertising budgets, payment for content creation services, fee for graphics, etc.

10 Conclusion

Talent need for the successful marketing. Identify the personnel required with a certain level of knowledge and skills on various aspects of channels. It also identify that is there sufficient manpower resources available or need to outsource. Various skills comprise development skills, creative skills, search skills, and social skills. Don't follow single formula to generate the result marketer want. All digital marketing channels have strength and weakness both. Thus, use of integrated channels is best way to reach end goals. Digital marketing is the effective technique for any business in any industry. Company identifies their audience online and started targeting them. Company identifies their needs, and create value for them. But it may not be effective to use same technique digital marketing strategy in all the business whether it is B2B or B2C. The buyer & seller tradeoff is also valuable.

Before creating content for the digital marketing first answer the above questions. To create the content for the digital marketing first step is to identify the need of the audience. By creating free templates or makemypersona.com marketers identify the audiences need in relation to the business. Identifying the need at various stages of the buyer's journey helps the marketers to reach their goals. This is known as content mapping (Yesbeck Jennifer). Digital marketing is the need of professionals. Marketers should have to plan in advance their marketing strategies and implement it by using digital media channels.

References

Aguila, M. (2019). *7 digital marketing channels*. https://blog.spiralytics.com/digital-marketing-channels.

Bullock, L. (2019). https://www.smartinsights.com/digital-marketing-strategy/create-an-effective-digital-marketing-budget-2020/.

Chaffey, D. (2017). *Digital marketing strategy advice—Smart insights digital marketing [online] smart insights*. http://www.smartinsights.com.

Chaffey, D., & Ellis-Chadwick, F. (2012). *Digital marketing: Strategy, implementation and practice* (1st ed.). Harlow: Pearson Education.

Das, S. (2020). Innovations in digital banking service brand equity and millennial consumerism. *Digital transformation and innovative services for business and learning* (pp. 62–79). IGI Global. https://doi.org/10.4018/978-1-7998-5175-2.ch004.

Das, S. (2021). *Search engine optimization and marketing: A recipe for success in digital marketing* (1st ed.). CRC Press. https://doi.org/10.1201/9780429298509.

Das, S., & Mondal, S. (2016). *Yours digitally: Patanjali*. Available at SSRN 2737299.

Das, S., Mondal, S. R., Sahoo, K. K., Nayyar, A., & Musunuru, K. (2018). Study on impact of socioeconomic make up of Facebook users on purchasing behavior. *Revista Espacios, 39,* 28–42. Retrieved from http://www.revistaespacios.com/a18v39n33/18393328.html.

Das, S., & Nayyar, A. (2019). Innovative ideas to manage urban traffic congestion in cognitive cities. In *Driving the development, management, and sustainability of cognitive cities* (pp. 139–162). IGI Global. https://doi.org/10.4018/978-1-5225-8085-0.ch006.

Das, S., & Nayyar, A. (2020). Effect of consumer green behavior perspective on green unwavering across various retail configurations. In *Green marketing as a positive driver toward business sustainability* (pp. 96–124). IGI Global. https://doi.org/10.4018/978-1-5225-9558-8.ch005.

Das, S., Nayyar, A., & Singh, I. (2019). An assessment of forerunners for customer loyalty in the selected financial sector by SEM approach toward their effect on business. *Data Technologies and Applications, 53*(4), 546–561. https://doi.org/10.1108/DTA-04-2019-0059.

Dodson, I. (2016). *The art of digital marketing* (1st ed.). Hoboken, NJ: Wiley.

Duy, N. T., Mondal, S. R., Van, N. T. T., Dzung, P. T., Minh, D. X. H., & Das, S. (2020). A study on the role of web 4.0 and 5.0 in the sustainable tourism ecosystem of Ho Chi Minh City, Vietnam. *Sustainability, 12*(17), 7140. https://doi.org/10.3390/su12177140.

Effectiveness of Online Digital Media Advertising as A Strategic Tool for Building Brand Sustainability: Evidence from FMCGs and Services Sectors of Pakistan. (2019). *Sustainability, 11,* 3436. https://doi.org/10.3390/su11123436.

French, A., & Smith, G. (2013). Measuring brand association strength: A consumer based brand equity approach. *European Journal of Marketing, 47*(8), 1356–1367.

Gupta, D. K., Jena, D., Samantaray, A. K., & Das, S. (2019). HRD climate in selected public sector banks in India. *Revista ESPACIOS, 40*(11).

https://digitalmarketingphilippines.com/how-to-choose-the-best-digital-marketing-channel-for-your-business/.

https://ecommercetrainingacademy.com/most-important-digital-marketing-channels-for-ecommerce/.

https://masterful-marketing.com/marketing-communications-channel-strategy/.

https://www.statista.com/statistics/304827/number-of-facebook-users-in-india/.

https://www.towermarketing.net/blog/digital-marketing-mix-4ps/.

Jain, S., Jain V., & Das, S. (2018). Relationship analysis between emotional intelligence and service quality with special evidences from Indian banking sector. *Revista ESPACIOS, 39*(33).

JENNIFER. https://blog.alexa.com/content-mapping/.

Lin, C., & Rauschnabel, P. A. (2015). Social media marketing: Psychological insights, managerial importance, and future research directions. In L.

Inn (Ed.), *Encyclopedia of E-commerce development, implementation, and management*. IGI Global (Forthcoming).

Mohanty, P. C., Dash, M., Dash, M., & Das, S. (2019). A study on factors influencing training effectiveness. *Revista Espacios, 40,* 7–15. Retrieved from http://www.revistaespacios.com/a19v40n02/19400207.html.

Mondal, S. R. (2020). A systematic study for digital innovation in management education: An integrated approach towards problem-based learning in Vietnam. In *Digital innovations for customer engagement, management, and organizational improvement* (pp. 104–120). IGI Global. https://doi.org/10.4018/978-1-7998-5171-4.ch006.

Mondal, S., Das, S., Musunuru, K., & Dash, M. (2017). Study on the factors affecting customer purchase activity in retail stores by confirmatory factor analysis. *Revista Espacios*, *38*(61).

Mondal, S., Mall, M., Mishra, U. S., & Sahoo, K. (2017). Investigating the factors affecting customer purchase activity in retail stores. *Revista Espacios*, *38*(57).

Mondal, S., & Sahoo, K. K. (2020). A study of green building prospects on sustainable management decision making. In *Green building management and smart automation* (pp. 220–234). IGI Global. https://doi.org/10.4018/978-1-5225-9754-4.ch011.

Mulhern, F. (2009). Integrated marketing communications: From media channels to digital connectivity. *Journal of Marketing Communications, 15*(2–3), 85–101. https://doi.org/10.1080/13527260902757506.

Nadanyiova, M., & Das, S. (2020). Millennials as a target segment of socially responsible communication within the business strategy. *Littera Scripta, 13*(1), 119–134. https://doi.org/10.36708/Littera_Scripta2020/1/8.

Patlin. (2019). https://patlin.com.my/2019/04/08/differences-digital-marketing-online-marketing/.

Ryan, D. (2014). *Understanding digital marketing: Marketing strategies for engaging the digital generation Ed. 3* (1st ed). London: Kogan Page.

Shamsudeen Ibrahim, S. A., & Ganeshbabu, P. (2018). A study on impact of social media marketing trends on digital marketing. *Shanlax International Journal of Management, 6*(1), 120–125. https://doi.org/10.5281/zenodo.1461321.

Sharma, E., & Das, S. (2020). Measuring impact of Indian ports on environment and effectiveness of remedial measures towards environmental pollution. *International Journal of Environment and Waste Management, 25*(3), 356–380. https://doi.org/10.1504/IJEWM.2019.10021787.

Singh, S., & Das, S. (2018). Impact of post-merger and acquisition activities on the financial performance of banks: A study of Indian private sector and public sector banks. *Revista Espacios Magazine, 39*(26), 25.

Singh, L. B., Mondal, S. R., & Das, S. (2020). Human resource practices & their observed significance for Indian SMEs. *Revista ESPACIOS, 41*(7). Retrieved from http://www.revistaespacios.com/a20v41n07/20410715.html.

Singh, S., Mondal, S., Singh, L. B., Sahoo, K. K., & Das, S. (2020). An empirical evidence study of consumer perception and socioeconomic profiles for digital stores in Vietnam. *Sustainability, 12*(5), 1716.

Singh, I., Nayyar, A., & Das, S. (2019). A study of antecedents of customer loyalty in banking & insurance sector and their impact on business performance. *Revista ESPACIOS, 40*(6).

Singh, I., Nayyar, A., Le, D.H., & Das, S. (2019). A conceptual analysis of internet banking users' perceptions: An Indian perceptive. *Revista ESPACIOS, 40*(14).

Talafuse, A. (2014). Traditional and digital media advertising preference: A descriptive evaluation of billings, Montana mothers. *Journal of Management and Marketing Research, 19*, 1–22.

Van, N. T. T., Vrana, V., Duy, N. T., Minh, D. X. H., Dzung, P. T., Mondal, S. R., & Das, S. (2020). The role of human–machine interactive devices for post-COVID-19 innovative sustainable tourism in Ho Chi Minh City, Vietnam. *Sustainability, 12*(22), 9523. https://doi.org/10.3390/su12229523.

CHAPTER 10

A Review for Managerial Guidelines for Social Media Integration of IMC in Digital Era

Bikramjit Rishi and Neha Reddy Kuthuru

1 Introduction

Organizations, for years, have advocated the importance of integration of various business activities to ensure consistency and cohesiveness in executing business goals (Singh, Mondal, Singh, Sahoo, & Das, 2020). With the growing popularity of integrating marketing activities, a significant amount of research has been undertaken by the researchers to understand the scope and applicability of Integrated Marketing Communication (IMC) (Anand, 2020; Finne & Grönroos, 2017; Vollero, Schultz, & Siano, 2019).

Conrnelissen and Lock (2000) discuss the emergence of IMC to be instead natural, citing that the integrated practices were being undertaken long before the development of its academic definition. It highlights the

B. Rishi (✉)
Institute of Management Technology (IMT), Ghaziabad, India
e-mail: brishi@imt.edu

N. R. Kuthuru
Shiv Nadar University, Greater Noida, India

S. Das and S. Gochhait (eds.), *Digital Entertainment*,
https://doi.org/10.1007/978-981-15-9724-4_10

fact that the IMC processes are inevitable for an organization (Singh, Mondal, & Das, 2020) and become the core element in achieving various business objectives.

Smith, Berry, and Pulford (1999) proposed that the integration of various marketing activities happens at the following levels:

1. Vertical integration of objectives—This level involves the alignment of marketing objectives to the organization's broader goals and mission (Sharma & Das, 2020).
2. Functional/horizontal integration—The organization tries to unify the different activities of the business functions like human resource management, operations, production with marketing activities (Das, 2020).
3. Integration of the marketing mix—The organization makes an effort to align the decisions regarding product, and price should be in sync with the place and promotion decisions (Mondal, 2020).
4. Integration of communication mix—The communication mix integration in the buyer journey. Consistency in the communication message across the buying journey process.
5. Integration of creative design—An effort is made by the organization to bring uniformity and consistency in the process of design. This effort helps in aligning the process with the positioning of the product (Mondal & Sahoo, 2020).
6. External/Internal integration—This level involves the cohesion between the various internal departments and the external teams hired to achieve a mutual objective (Das & Nayyar, 2020)
7. Financial integration—Ensuring that the execution of marketing activities is as per the relevant budgets and is economical and profitable for the organization (Das, Nayyar, & Singh, 2019).

Hence, IMC achieves the objectives as mentioned above by integrating different elements of marketing communication mix like sales promotion, advertising, public relations, personal selling, and direct marketing (Das & Nayyar, 2019). On this basis, Marketers can understand the benefits of IMC on two different fronts. From the customer's end, they are provided with all the relevant information consistently to make an informed decision Singh, Nayyar, Le, & Das, 2019) & on the organization's side, they can achieve a consolidated brand image while ensuring smooth execution of business processes (Mohanty, Dash, Dash, & Das, 2019).

IMC, like any other business process, is a function of specific factors that are bound to change across the lifespan of the organization (Singh, Nayyar, & Das, 2019). Technological development has noticeably accelerated numerous business processes from production to delivery (Gupta, Jena, Samantaray, & Das, 2019). It has also led to significant marketing developments and diffusion of IMC. With the increasing penetration of the Internet, coupled with the growing affordability of broadband connections, has played an essential role in changing the landscape for marketing activities (Singh & Das, 2018). With the rise of the Internet and the emergence of various social media platforms, there has been a dramatic change in the way consumers consume and how the organizations approach their marketing strategy (Jain, Jain & Das, 2018). These rapid technological advancements have led to the development of a novel subset of marketing strategy, which is called Digital Marketing (Das, Mondal, Sahoo, Nayyar, & Musunuru, 2018).

2 Digital Marketing and IMC

Kannan and Li (2017) defined Digital Marketing as "an adaptive, technology-enabled process by which firms collaborate with customers and partners to jointly create, communicate, deliver, and sustain value for all stakeholders." The technological developments and the digital transformation have, in a way, brought the businesses closer to its audience and provided a platform to engage in a dialogue in the most organic way possible (Mondal, Das, Musunuru, & Dash, 2017). The organizations are increasingly adopting revolutionary technologies like Artificial Intelligence, Machine Learning, and several such technologies to gain a deep understanding of consumer behavior to gain insights that could be integrated by the marketers into the development of marketing strategy (Das, 2021; Duy et al., 2020; Mondal, Mall, Mishra, & Sahoo, 2017).

Makrides (2017) reported that with the increasing internet consumption, there had been a shift in the power dynamic from the companies to the consumers, with the latter assuming significantly higher power and controlling the narrative of the brand. Lamberton and Stephen (2016), through thematic analysis, explained the Internet as a tool that can be effectively used by the marketers for Marketing Intelligence. They highlighted the use of the Internet to study the preferences of the consumers to provide them with the most relevant products, which would aid them in the purchase process and develop a close association with the brand (Das & Mondal, 2016). Makrides, Vrontis, and Christofi (2019) defines digital marketing as a tool to enable two-way communication with consumers to drive brand loyalty (Van et al., 2020).

The digital transformation has also made the competition in the marketplace increasingly aggressive. The ease of access to digital marketing resources and the convenience involved in approaching the customers has further intensified the competition. While this increases the pressure on companies to gain a competitive advantage, it also provides a high degree of value to small and medium-sized businesses who can now develop a relationship with the buyers at a little cost. Advertising forms a core part of any IMC strategy by communicating the information about the brand/product to the relevant target audience. With a significant number of buyers using digital media to gather information about multiple purchases, marketers must direct their advertising efforts through suitable digital channels. Rosengren (2019) discussed extensively advertising in the digital media landscape by highlighting the growing importance of customer-centered communication. With the increasing popularity of blogs and social media platforms, consumers are acting as co-creators in the campaigns and being the face of the brand for their advertising.

Makrides et al. (2019) emphasized the importance of word-of-mouth marketing, which has a considerable influence on the purchase decisions of the consumers. Digital marketing has thus paved the way to the development of another mode of marketing called Influencer Marketing, which in itself captures the essence of word-of-mouth advertising in its most real sense. Digital influencers are opinion leaders on digital platforms who are either experts in a particular domain or someone who has numerous social connections (Goldenberg, Lehmann, Shidlovski, & Barak, 2006). Influencers possess expertise about a specific product or category. The followers consider their recommendation when making a purchasing decision. Choi and Rifon (2012) explain that the perceived influence of these opinion leaders is more substantial when the consumers' personality resonates with the kind of content published by the influencers. It has a direct implication on the Marketers who have to make a conscious choice when onboarding influencers. They have to choose people whose style and personality align with the brand's identity and image. Additionally, it is equally essential for the content created by the influencers to be original and unique to have a significant impact on the viewers (Casaló, Flavián, & Ibáñez-Sánchez, 2018).

2.1 *Impact of Digital Consumers on Digital Marketing*

In simple terms, Digital Consumers are the ones who use the Internet and digital platforms to buy or sell products. Digital consumers also leverage the use of digital platforms to share their experiences with the brands. In a way, consumers have taken control of the narrative of the brand across various digital platforms (Tuten, Solomon, & Rishi, 2020). The information-seeking and decision-making process has transitioned to an online medium with consumers looking for details, recommendations, and reviews on an online platform. Thus, it becomes imperative for Marketers to have control over the brand communication across different online channels to maintain consistency and authenticity. With the most brands increasingly scaling up their digital marketing initiatives, the consumer is getting exposure to hundreds of brands, and the brands must gain the top of the mind awareness. It is possible when the brands maintain a strong relationship with the consumers and engage with them on these digital platforms in the most organic way possible.

Tkaczyk (2016) discusses the digital consumer in detail and emphasizes that digital consumers want to be a co-creator in curating the messaging and demand to be an active recipient of the brand's communication. The importance of personalizing the digital experience is increasing due to the online interactions with the consumers. Hence, as digital age Marketers, they should have a sound understanding of the consumer to maintain a strong relationship with them (the consumers), which would translate into improved brand awareness and, ultimately, higher sales.

3 Digital First Approach & Perspectives of Social Media

A Forbes report reported that 55% of the global population has access to the Internet, and the users execute about 51% of the digital transactions with the help of mobile devices (Evans, 2020). The "Digital Advertising in India 2020" report by Dentsu Aegis Network valued the digital advertising industry in India at INR 13683 crores, with the sector expected to grow at a CAGR of 27.42%. The same report identified Consumer

Durables, E-Commerce, Telecom, Banking, and Financial Service sectors to be the largest spenders for digital advertising.

With this significant increase in the usage of digital platforms, it becomes imperative for the organizations to look at *digital* as one of their primary communication mediums to stay relevant and reach their target audience. Since the consumers are moving toward an online form of interaction, it is sensible that the organizations prefer digital channels over traditional media to release new content.

A digital-first approach toward marketing provides an advantage over traditional media like print, TV in a way that the former is way more responsive to customer interaction. Organizations can understand in real time how receptive the customers are of a particular campaign and can make necessary changes to position the product effectively. The digital platforms are dynamic and provide an understanding of the changing needs and preferences of the consumers. It assists the brands in developing a strong and personal relationship with the audience. Hence, it is essential to analyze the elements of digital marketing and understand the process of its integration in a marketing strategy.

4 Introduction to Social Media Marketing

Nair (2011) explained that social media is a platform to share content, opinions, insights, and media. He further emphasized that social media, at its core, is centered around the relationships/connections between people and organizations. Social media advertising spending is forecasted to increase by about 20% to 43 billion dollars in 2020 ("US Social Trends for 2020," 2020). It is evidence enough that it is essential to consider social media as an integral, if not the most critical part of a company's digital strategy.

Social media platforms thus emerged as effective channels of communication that facilitate the organizations in reaching their target customers. With the increasing number of social media platforms, it becomes essential to understand the characteristics of the different platforms to choose the one that best matches the objectives of the company.

4.1 Facebook

As of the first quarter of 2020, Facebook has a whopping 2.6 billion active users every month. It makes Facebook the most significant social network worldwide (Clement, 2020). Over the past decade, due to its phenomenal popularity, Facebook became one of the sorts after marketing platforms for companies worldwide. The platform offers an opportunity to share content in text, picture, or video format with your connections and react to the posts posted by the contacts.

Facebook also provides an opportunity to run paid marketing content. Upon running a paid marketing campaign, the information reaches the audience through an in-built algorithm. The Facebook algorithm shows the most relevant kind of posts based on a user profile, interests, and search history. Hence, it is very crucial for Marketers to carefully align their communication messages on Facebook with the campaign objectives to reach the most relevant target audience. Impressive images and engaging videos that call for action from viewers are the most impactful approach toward executing a Facebook campaign. Facebook also provides free business tools and insights to track the impact of any campaign.

Due to access to customers worldwide, the platform can also be effectively used for Market Research purposes to understand the customer trends and insights that are helpful in new product development and marketing. Facebook is also one of the best social media platforms for instant customer engagement. Information immediacy and immediate feedback from customers makes Facebook a great platform to develop strong relationships with customers.

4.2 Instagram

Instagram is a social media network based on media sharing and is an extension of Facebook. As of 2019, the platform has one billion people every month (Newberry, 2019). Instagram allows users to create photos, video content, and organically interact with their community. It also provides a platform for businesses to run sponsored posts to increase brand awareness, drive sales, or generate leads. The sponsored posts use location, demographics, interests, and behavior to target consumers. The platform also provides businesses with tools to track metrics like engagement and reach.

Another critical characteristic of Instagram is the popularity of influencer marketing on the platform. These influencers aim to build a community of people with similar likes and preferences. They maintain close-knit connections with this community by leveraging several in-built tools like Stories, Feed, IGTV, and Live. Influencer marketing is becoming increasingly popular with Marketers dedicating considerable budgets to promote their products using influencers. Influencer marketing provides the advantage of targeted word-of-mouth marketing by taking a tone of advocacy to increase brand awareness.

4.3 *YouTube*

Since its launch in 2005, YouTube has grown at an exceptional pace to become the most popular video streaming service worldwide. As of 2019, the platform had two billion logged-in monthly users (Clement, 2019). The popularity and the reach of YouTube present the organizations with an opportunity to grow their brands through interactive videos. The platform offers marketing opportunities on two fronts. The brands can develop their channel to provide information about the products or run sponsored video campaigns on other channels. It is crucial to understand the target audience's behavior before executing a YouTube campaign. Specific vital questions regarding the kind of videos watched by the target audience, the average watch time, the demographics help in while planning a YouTube campaign. The platform also provides an opportunity to understand customer reactions and feedback to a YouTube campaign. It helps in making suitable changes to the campaign.

4.4 *Twitter*

Twitter is also one of the most popular social media platforms worldwide, with about 330 million monthly active users (Lin, 2020). Twitter has, over the years, emerged into a reliable marketing platform that lets the brands interact with their customers in real time. On similar lines with Instagram and Facebook, Twitter also provides an option to run paid ad campaigns on the platform. The sponsored tweets appear on the timeline of users that share similar interests with the target audience.

Unlike other social media platforms, Twitter limits its users to make posts using just 280 characters. It means that the marketers have to use the limited space in the most creative way possible to attract the attention of the viewers. Hashtags and trends help in increasing engagement with the users. The essence of a successful Twitter campaign lies in building a two-way communication with the audience and creating an engaging dialogue with them.

4.5 *LinkedIn*

With over 675 million monthly users, LinkedIn focuses on building professional networks and connections (Rishi & Bandyopadhyay 2017). It provides users the opportunity to connect with professionals with the same interests and expand their network. Organizations can set-up their company page on the platform to present their followers with information regularly. LinkedIn can be used to create brand awareness, scout for talent, and generate sales leads.

LinkedIn is unique in comparison to the other social media platforms, as it has a more professional tone. It provides a platform to engage with professionals from across the globe (Nadanyiova & Das, 2020). Thus it can be an excellent platform to generate B2B connections and leads. The platform includes business tools like sponsored content, sponsored messaging, text, and dynamic ads. The marketers can identify the right kind of business tool based on the campaign objective and the target audience. The platform also provides an analytics tool to track the impact of a campaign, the number of leads generated, the number of visitors.

5 Social Media and Marketing Strategy

Now that we have established the characteristics of different social media platforms, it is essential to discuss in detail the importance of integrating the appropriate platform into the overall marketing strategy of a brand. Research has shown that marketers are increasingly leveraging social media networks to communicate with their customers. Marketing efforts on social media platforms encourage online interactions between the brand and consumers and further strengthen the relationship between them (Murdough, 2009). Irrespective of the marketing objectives, it is crucial for the content created by the brand to be relevant to the customers for them to engage in the most organic way possible (Schmitt, 2012).

Following themes can help in understanding the importance of social media marketing and incorporating it into the marketing strategy:

5.1 *Understanding the consumer and Social Listening*

Due to the sheer number of users on social media platforms, brands can use these platforms to understand consumer needs and preferences. Observing consumer interactions on these platforms would give the marketers insight into consumer behavior. The emergence of social media platforms has elevated the significance of the process of *Social Listening*. Stewart and Arnold (2018) defined social listening as "an active process of attending to, observing, interpreting, and responding to a variety of stimuli through mediated, electronic, and social channels." Social listening has a direct implication on strategy by gaining a deep understanding of consumer needs and experiences.

A deep dive into the social media activity of the target audience helps in understanding the perceptions of the brand and the reactions to various products, which could either be positive or negative. Due to the immediacy in response to social media, marketers can incorporate the relevant insights and feedback to enhance the features of the product. Social media also provides a great platform to address customer queries and concerns almost immediately. It would lead to enhanced customer experience and increased customer satisfaction. Hence, through the process of social listening on social media platforms, brands can create products that match the customer needs/expectations, study the industry trends and enhance the relationship with the customers by engaging with them regularly.

5.2 *Brand Awareness and Advocacy*

Effective use of social media resources can significantly increase the brand/product awareness for a company. Through various social media initiatives, consumers get exposure to different elements of the brand, which provides relevant information and eventually increases the recognition of the brand. Stephen and Galak (2012) discussed that the social media interactions like referring/tagging a friend, or the discussions that happen in the online communities/pages play a significant role in influencing purchase decisions, new customer acquisition, and sales. Customers often rely on opinions and recommendations of people who have already used the product/brands to make their own purchase decisions. Thus, social media provides access to product information and reviews that assist in making an informed purchase decision. Wang, Yu and Wei (2012) discuss that peer communication through social media platforms has a positive influence on the purchase intention of consumers. This influence is found to happen in two ways—a direct impact through conformity and an indirect influence by strengthening product involvement. By creating an association and engagement with the product/brand, the marketers can effectively acquire new customers and retain the existing customers. Social media marketing also has a direct impact on brand loyalty, and the effect is higher when the customers see interactive campaigns and relevant content (Erdoğmuş and Çiçek, 2012).

5.3 Generating Leads and Driving Sales

Through various social media platforms, the brands can drive campaigns that boost engagement and significantly increase the number of leads generated. Marketers are increasingly undertaking creative initiatives to track user-generated content. It implies that the viewers produce organic content on social media by sharing the campaigns created by the brand. User-generated content is an effective way of tracking customers who are interested in the brand/product. The brands can then target their marketing toward these potential customers to convert them into customers. Personalization of the content by using brand/product-specific hashtags are one way to track the user-generated content.

Strong engagement is the key to acquiring customers on social media. Some proven ways of gaining a strong commitment are to host engaging sale promotion contests that drive user-generated content. Social media makes the process of driving sales more efficient in a way that the brands can initiate a call to action, which forces the viewers to the website to make a purchase decision.

5.4 Increasing ROI (Return on Investment)

Social media has two types of impacts. The first being the qualitative impact that it creates in the form of increased brand awareness, consumer engagement, or loyalty. The other is the quantifiable impact in the way of increased web traffic, sales. The qualitative result is not revenue centric, but it provides clear insights about the brand and the campaign. To measure the quantifiable impact, marketers can define specific KPIs (Key Performance Indicators) to track the increase in revenue, leads generated, etc. The marketers can calculate the ROI by comparing the outcomes with the objectives of the campaign. Kumar and Mirchandani (2012), through their research, concluded that the companies could improve the ROI of its social media campaigns by identifying and incentivizing influencers on these platforms to talk about the products or services.

A report suggests that about 48% of marketers agree that they have seen a measurable ROI through social media marketing (Allen, 2017). Influencer marketing, which is a subset of social media marketing, has also proven to show great results. According to the State of Influencer Marketing 2020 Report, for every $1 spent on influencer marketing, the brands earned about $5.78 in Average Earned Media Value ("The State of Influencer Marketing 2020: Benchmark Report," 2020). Earned media value aims to capture the promotion efforts, excluding the paid campaigns. It captures the brand awareness generated through endorsements by influencers. This high ROI through social media channels shows that companies can achieve their marketing objectives with minimal spending and in a cost-effective way.

6 Social Media Strategy, Planning, and Execution

The following steps can help in understanding the process of planning & development, and the execution of any social media campaign (Tuten et al. 2020).

a. Setting objectives

The first step in planning a social media strategy is to identify objectives for running a campaign. Defining clear social media objectives enables social media managers in effective planning and execution. Critical decisions regarding the target audience, choice of platform, budget depend directly on the social media objectives of the campaign. The applications of social media marketing can be in understanding the purchase behavior (Relling, Schnittka, Sattler, & Johnen, 2016), customer relationship management (Trainor, Andzulis, Rapp, & Agnihotri, 2014), and brand management (Asmussen, Harridge-March, Occhiocupo, & Farquhar, 2013). Based on these applications, some objectives are—Increasing brand awareness, customer engagement (Ashley & Tuten, 2015), improving sales, generating web traffic, improving brand image, and reducing marketing costs (Felix, Rauschnabel, & Hinsch, 2017).

b. Know your consumer

It is one of the critical processes in the planning and execution of a social media strategy. The success of any campaign depends on how effectively the brands can communicate to their target audience and achieve the objectives. Hence it becomes imperative to have an in-depth understanding of the characteristics of the target consumers. Through the process of social listening, analysis of consumer behavior helps in marketing decisions (Stewart & Arnold, 2018). This process has relatively become more straightforward due to the readily available demographic data and access to various analytics tools. Social media demographics data gives an exhaustive understanding of which platforms the brands should approach and the kind of content that has to be created by the brand. Specific other metrics like the average time spent on the social media platforms, the type of content consumed by the followers, and the interactions on the platforms give the marketers a further insight to formulate the social media strategy. The brands should also develop buyer personas that capture the needs and preferences of the target consumer. It would assist in developing the most appropriate communication message and curating engaging content.

c. Choice of Platform

The available social media platforms differ in the type of content, the tone, user personas, etc. Social media managers need to identify a suitable platform that allows them to achieve their business goals effectively. The characteristics and benefits of the different platforms should be understood and matched to the objectives of a campaign. Recognizing the ideal platform helps the brands in reaching their target audience cost-effectively and seamlessly. The target audience plays a significant role in deciding the social media platform. Marketers should research to identify the most used platform by the target audience. For example, if a brand is a B2B service agency, an ideal platform for undertaking social media initiatives would be LinkedIn owing to the extensive network of industry professionals.

d. Competitor Analysis

Upon choosing the ideal social media platform, marketers can undertake a social media audit of the competitors, present on the platform. Social media audit analyzes the relevant competitors for the review. Social media audit helps in deciding the type of content concerning the product category and the kind of engagement which needs to be managed by the marketers with the consumers (He et al., 2016). Competitor Analysis thus helps in designing the social media strategy.

e. Content creation and engagement

Creating engaging and relevant content is the key to a successful social media strategy. The content creation should be aligned with the business objectives and based on the understanding of the chosen social media platform and the target audience. Engaging and creative content makes the brand stand out on these platforms and gain a competitive advantage (Ashley & Tuten, 2015). It is also essential to define a brand voice and personality that remains consistent across various content posts. Since social media is primarily a visual medium, brands should create visually appealing and engaging content that aligns with the brand identity. Prompting user-generated content and by leveraging the use of hashtags, the brands can further increase the engagement. It is all the more important to focus on consumer engagement as engagement on online platforms translates into increased brand loyalty (Dessart, Veloutsou, & Morgan-Thomas, 2015).

f. Brand-consumer synergy

Social media provides a platform that drives organic interactions between the brands and the consumers. These interactions must create a positive impact on the consumers and influence them in making a purchase decision. It is also in line with the self-congruence theory that suggests the extent of similarity between the consumer's self-image and the image of the brand. A positive correlation between the two leads to positive behavior toward the brand and the contrary is applicable, too (Sirgy, 1982). Hence, to foster a strong relationship with the consumers, the brands need to understand the personality of its target audience and accordingly define their brand personality on social media. It is also important to consistently maintain the defined brand personality across the various content pieces on the platform. It ensures uniformity in communication and helps the consumers to identify the brand.

g. Metrics and Analytics

The last step in executing a successful social media strategy is to measure the results of the various initiatives. It becomes imperative to identify and define relevant key performance indicators for the evaluation of a campaign (Keegan & Rowley, 2017). Defining the relevant metrics helps the brands in optimizing their strategy to achieve long-lasting results. Some critical parameters for consideration are—awareness, engagement, and ROI. Kurniawati, Shanks, and Bekmamedova (2013) explain the importance of social media analytics in terms of improved marketing strategy and better customer engagement. The analyzed results have to be then related to the objectives that were defined earlier to understand the impact of the marketing efforts.

7 Managerial Toolkit for Social Media Integration

Implementation of a successful social media strategy can be done most efficiently by leveraging the use of the following tools:

a. Brand Guidelines

Brands have a personality of their own that defines the brand and forms the basis for consumers to relate to (Aaker, 1997). Consistency is the key to building a memorable brand, and a set of Brand Guidelines helps the social media team with this. Brand guidelines comprise a mission statement, user personas, tone of voice, color schemes, the brand story, and description of the brand personality. This kind of formal report guides the team in curating social media content that aligns with the brand identity and business goals (Modicum, 2016).

b. Social Media Calendar

Planning is key to developing a creative social media strategy, and having a single platform to manage all the content would be highly efficient for the social media team. Social media calendars provide a highly accessible way of planning and tracking social media content. It would further help in the alignment of the marketing strategies with the overarching objectives to ensure consistency in communication. It would also assist in scheduling the various content pieces to when they are most relevant to and track the feedback received from the audience. Marketers should incorporate in refining the strategy to achieve the maximum results. Social media calendars also facilitate collaboration among various departments involved in the process (Aboulhosn, 2020).

c. Analytics Tools and Dashboards

Social media analytics play a vital role in analyzing the impact of the executed social media campaigns. The data-driven approach used in analytics helps in making social media decisions more successful. Every Marketer has to define relevant metrics that are in line with social media marketing objectives. A dashboard provides the benefit of managing analytics and metrics for various content pieces on a single platform. It provides a collaborative platform that helps in monitoring and tracking social media content. It also assists in generating in-depth insights about the engagement with the target audience and the effectiveness of executed strategies (Murdough, 2009).

8 Challenges in Social Media Integration

While social media presents a plethora of opportunities for executing marketing campaigns, it also has its share of challenges. Hofacker and Belanche (2016) identified customer creation and engagement as one of the problems of social media marketing. With the increasing importance of customer engagement and fostering a strong relationship with them, it becomes challenging to personalize the communication message. Marketers can overcome this challenge by leveraging the use of storytelling and taking a humanistic approach toward communication. Emphasis should be on delivering value instead of selling the products directly. Brands should also invest in developing creative ideas to prompt engagement and user-generated content.

Another challenge faced by social media managers is to be able to measure the impact of a campaign objectively. Since the interactions taking place on the platforms, and the awareness created is qualitative, marketers find it difficult to assess the ROI on social media. We can overcome this by identifying the relevant key performance indicators and metrics for the brand. It is also essential to realign the social media objectives to the broader business goals.

Yadav (2017) addressed the challenges for privacy and security risks. The rising popularity of social media also makes it sensitive to security breaches. Since constant information transfer takes place on these platforms, transparency has to be maintained with the customers to gain their trust. Organizations have to be hence conscious about their activities on social media and continuously monitor it to avoid any cases of a security breach or brand impersonation. By addressing the challenges, social media can become a potent tool that can be leveraged for marketing to create a substantial impact on business goals.

9 Case Studies

With the growing popularity of social media, several brands benefited from these platforms in terms of increased sales, awareness, and loyalty. Case studies serve as a tool in understanding how the brands have successfully implemented their social media strategy and draw inspiration from it. The following case studies are examples of how brands from different sectors accelerated their marketing efforts through social media while gaining impactful results.

Dove has pioneered the art of social media marketing that builds on a strong understanding of consumer insights. Most content created by Dove on social media focuses on the basic premise of going beyond and breaking the stereotypical standards of beauty. Their research showed that about 80% of women face negativity on social media (Shorty Awards, 2016). These negative conversations have a substantial impact on the confidence of women. To bring about a change in the beauty conversations that happen on social media, Dove launched their #SpeakBeautiful campaign. The campaign targeted toward raising awareness about body shaming on social media to foster confidence in women concerning how beauty is perceived. The company has chosen Twitter as the platform for the execution of this campaign. The hashtag encapsulated the idea of the campaign in its most accurate form and prompted the creation of user-generated content in terms of making positive tweets. Dove, in collaboration with Twitter, also built a tool to assess the positivity of a particular tweet (Johnson, 2016). The campaign has gained phenomenal reach with hashtags has been used over 168,000 times while the campaign gained about 800 million impressions (Shorty Awards, 2016). This campaign is a classic example of understanding the consumer and undertaking initiatives to drive engagement among the target audience.

Airbnb, an online marketplace for rental accommodations, is another brand that has successfully integrated social media into its business strategy. Through their social media initiatives, Airbnb was able to create a global online community based on the values of hospitality to build brand awareness and acquire new customers (Harris, 2015). The #OneLessStranger campaign used the consumer insight that most customers were uncomfortable with living in a stranger's home. Through this campaign, they aimed to bring strangers closer to social media by using pictures, stories, and acts of kindness to inspire one another (Spary, 2020). The outcome of the campaign was increased brand awareness, with more customers being made aware of the concept of Airbnb's business. The campaign was successful in building trust among prospective users. This campaign shows how effective user-generated content can be for increasing brand awareness on social media.

Influencer marketing is a subset of social media marketing that has recently been getting popular with its highly effective results. Dunkin Donuts has used the influencer marketing tool to raise awareness for their handcrafted espresso drinks. They got onboard millennial Nano and micro-influencers who have about 1000–10000 followers on Instagram in an attempt to connect with the millennials. The tagline of the campaign was "Sipping is believing." The objective of the campaign was to increase the engagement of the brand on the platform. The brand message was consistent, and the company has used a natural, authentic tone of voice. It was in line with the kind of audience they were targeting who belonged to the millennial group who prefer relatable content. The campaign was highly successful for the brand by focusing over a million followers and gaining an engagement rate of 5.2% (Mediakix, 2019). This campaign provides fundamental learning in terms of using influencers, whose image matches the brand image. It increases the reliability for the users, which in turn has an impact on awareness and engagement.

10 Conclusion and Future Scope

Despite being a relatively new tool for marketing, social media has grown and found itself as being a core element of any brand's marketing strategy. Now having a social media presence is not enough. The brands have to increasingly undertake initiatives to engage with the customers and strengthen their brand image. According to the "State of Marketing Report" by Salesforce, about 75% of the marketing leaders have seen a direct return on their social media investments (Salesforce, 2018). Backed by significant impact, social media marketing is just going to be adopted by more and more marketers. The following themes explain the future of social media marketing and the road ahead for marketers:

a. Rising Influencer Marketing

With the rise in the emergence of influencers on all social media platforms, the millennial and the Gen-Z users are embracing influencers. Brands need to adopt influencer marketing strategies as a part of their social media initiatives due to the low cost and high ROI generated. Influencers are adopting creative content strategies to increase their engagement further. In the future, micro and nano influencers who are backed by their niche play a significant role due to their highly engaging audience. Consumers would demand authenticity in the kind of content. Hence, it would be advisable for the brands to give complete control over the creative process to the influencers since they have a strong understanding of their audience.

User-generated content would become increasingly powerful, with customers looking for resources to make their purchase decisions. To achieve the maximum results from influencer marketing, the marketers need to assess the alignment of the chosen influencer with the brand image and values. The brand's target audience and buyer personas should be looked at and matched with the influencer's target audience. Since influencer marketing is only going to gain momentum in the future, the brands must build a strong relationship with the influencers.

b. Emphasis on Content and Storytelling

Content is going to play a vital role in gaining competitive advantage across social media platforms. Consumers will engage with more authentic content as compared to blatant advertisements. Thus, marketers have to make significant investments in developing a content strategy that stands out and is also relatable at the same time.

Due to the increase in demand for humanized content, storytelling is going to play an essential role in crafting brand experiences on social media. Compelling storytelling encompasses a deep understanding of the consumers to form a strong relationship with them. Marketers can leverage these relationships to increase brand awareness and loyalty. Brands should look at embedding the product/service into their narrative that comes across as relatable and genuine. The focus should be on providing value to the customers and thereby elevating the customer experience.

c. Analytics-driven strategy

The focus on data-backed business decisions is increasing day by day. It is also very much applicable in the context of social media marketing. Social media analytics assist the marketers in understanding the kind of impact created by the various campaigns on the platform. Insights from the analytics thus generated would play a vital role in further tailoring the social media strategy. Marketers should hence define relevant metrics and continuously monitor them to understand the impact of the campaign. The use of analytics can help in understanding the target audience, determining and designing the target group, and analysis of any changes in the target group. Analytics can also help in understanding the type of interactions the followers are having in the online community. These interactions can act as a great source of consumer insights for designing the marketing strategy. These insights would reflect the brand perceptions among the consumers so the brands can deliver what their consumers need.

d. An Omni-Channel Strategy

Consumers these days use multiple channels and platforms to research about the product/service and make their purchase decisions. Hence, brands must have an omnichannel presence to keep up with their customer base. It involves developing synergy and integrating the different channels of communication used by the brands. Through this integration, brands can portray a consistent message while assisting in every stage of the buyer's journey. It also provides the advantage of understanding the changing user patterns across different platforms to provide a more personalized consumer experience. Social listening and market research will help in getting a better understanding of the purpose behind using a particular channel by the consumers. It would play a role in curating the content for that specific channel. For example—if research indicates that consumers are more likely to use Twitter to express their concerns, then a dedicated customer service team can be deployed to work on that channel. This kind of optimization creates a more holistic marketing experience, which is crucial to enhance the overall consumer experience.

The future of social media is highly exciting, and it has become a need of the hour to ensure the smooth integration of social media into the overall business strategy. Due to its high relevance in this digital age, it is vitally important for brands and marketers to understand social media and design a holistic plan to engage their customers better.

References

Aaker, J. L. (1997). Dimensions of brand personality. *Journal of Marketing Research, 34*(3), 347–356.

Aboulhosn, S. (2020). *How to create a social media calendar that works [Blog]*. Retrieved from https://sproutsocial.com/insights/social-media-calendar/.

Allen, R. (2017). *Does Social Media marketing actually generate ROI?* Retrieved from https://www.smartinsights.com/social-media-marketing/social-media-analytics/social-media-marketing-actually-generate-roi/.

Anand, S. R. M. (2020). Digital integrated marketing communication (dimc) and it's logical relationship with co-branding and brand values with reference to coke studio. *Academy of Marketing Studies Journal, 24*(1), 1–21.

Ashley, C., & Tuten, T. (2015). Creative strategies in social media marketing: An exploratory study of branded social content and consumer engagement. *Psychology & Marketing, 32*(1), 15–27.

Asmussen, B., Harridge-March, S., Occhiocupo, N., & Farquhar, J. (2013). The multi-layered nature of the internet-based democratization of brand management. *Journal of Business Research, 66*(9), 1473–1483.

Casaló, L., Flavián, C., & Ibáñez-Sánchez, S. (2018). Influencers on Instagram: Antecedents and consequences of opinion leadership. *Journal of Business Research, 117,* 510–519.

Choi, S. M., & Rifon, N. J. (2012). It is a match: The impact of congruence between celebrity image and consumer ideal self on endorsement effectiveness. *Psychology and Marketing, 29*(9), 639–650.

Clement, J. (2019). *Global logged-in YouTube viewers 2019.* Retrieved from https://www.statista.com/statistics/859829/logged-in-youtube-viewers-worldwide/.

Clement, J. (2020). *Facebook: Active users worldwide.* Retrieved from https://www.statista.com/statistics/264810/number-of-monthly-active-facebook-users-worldwide/.

Conrnelissen, J., & Lock, A. (2000). Theoretical concept or management fashion? *Journal of Advertising Research, 40,* 7–15. https://doi.org/10.2501/JAR-40-5-7-15.

Das, S. (2020). Innovations in digital banking service brand equity and Millennial consumerism. In *Digital transformation and innovative services for business and learning* (pp. 62–79). IGI Global. https://doi.org/10.4018/978-1-7998-5175-2.ch004.

Das, S. (2021). *Search engine optimization and marketing: A recipe for success in digital marketing* (1st ed.). CRC Press. https://doi.org/10.1201/9780429298509.

Das, S., & Mondal, S. (2016). *Yours digitally: Patanjali.* Available at SSRN 2737299.

Das, S., Mondal, S. R., Sahoo, K. K., Nayyar, A., & Musunuru, K. (2018). Study on Impact of socioeconomic make up of Facebook users on purchasing behavior. *Revista Espacios, 39*, 28–42. Retrieved from http://www.revistaespacios.com/a18v39n33/18393328.html.

Das, S., & Nayyar, A. (2019). Innovative ideas to manage urban traffic congestion in cognitive cities. In *Driving the development, management, and sustainability of cognitive cities* (pp. 139–162). IGI Global. https://doi.org/10.4018/978-1-5225-8085-0.ch006.

Das, S., & Nayyar, A. (2020). Effect of consumer green behavior perspective on green unwavering across various retail configurations. In *Green marketing as a positive driver toward business sustainability* (pp. 96–124). IGI Global. https://doi.org/10.4018/978-1-5225-9558-8.ch005.

Das, S., Nayyar, A., & Singh, I. (2019). An assessment of forerunners for customer loyalty in the selected financial sector by SEM approach toward their effect on business. *Data Technologies and Applications, 53*(4), 546–561. https://doi.org/10.1108/DTA-04-2019-0059.

Dessart, L., Veloutsou, C., & Morgan-Thomas, A. (2015). Consumer engagement in online brand communities: A social media perspective. *Journal of Product & Brand Management, 24*(1), 28–42.

Duy, N. T., Mondal, S. R., Van, N. T. T., Dzung, P. T., Minh, D. X. H., & Das, S. (2020). A study on the role of web 4.0 and 5.0 in the sustainable tourism ecosystem of Ho Chi Minh City, Vietnam. *Sustainability, 12*(17), 7140. https://doi.org/10.3390/su12177140.

Erdoğmuş, İ., & Çiçek, M. (2012). The impact of social media marketing on brand loyalty. *Procedia—Social and Behavioral Sciences, 58*, 1353–1360.

Evans, M. (2020). *5 stats you need to know about digital consumers in 2020*. Retrieved 5 June 2020, from https://www.forbes.com/sites/michelleevans1/2020/02/04/5-stats-you-need-to-know-about-digital-consumers-in-2020/.

Felix, R., Rauschnabel, P. A., & Hinsch, C. (2017). Elements of strategic social media marketing: A holistic framework. *Journal of Business Research, 70*, 118–126.

Finne, Å., & Grönroos, C. (2017). Communication-in-use: Customer-integrated marketing communication. *European Journal of Marketing, 51*(3), 445–463.

Goldenberg, J., Lehmann, D. R., Shidlovski, D., & Barak, M. M. (2006). *The role of expert versus social opinion leaders in new product adoption* (Marketing Science Institute Report No. 06-124).

Gupta, D. K., Jena, D., Samantaray, A. K., & Das, S. (2019). HRD climate in selected public sector banks in India. *Revista ESPACIOS, 40*(11).

Harris, R. (2015). *Airbnb seeks to rid of the world of strangers with new initiative*. Retrieved 18 June 2020, from http://marketingmag.ca/advertising/airbnb-seeks-to-rid-of-the-world-of-strangers-with-new-initiative-134082/.

He, W., Xu, G., Kim, Y., Dwivedi, R., Zhang, J., & Jeong, S. R. (2016). Competitive intelligence in social media Twitter: iPhone 6 vs. Galaxy S5. *Online Information Review, 40*(1), 42–61.

Hofacker, C., & Belanche, D. (2016). Eight social media challenges for marketing managers. *Spanish Journal of Marketing—ESIC, 20*(2), 73–80. https://doi.org/10.1016/j.sjme.2016.07.003.

Jain, S., Jain V., & Das, S. (2018). Relationship analysis between emotional intelligence and service quality with special evidences from Indian banking sector. *Revista ESPACIOS, 39*(33).

Johnson, L. (2016). *Dove and Twitter built a tool to measure how positive or negative your tweets are*. Adweek.com.

Kannan, P. K., & Li, H. A. (2017). Digital marketing: A framework, review and research agenda. *International Journal of Research in Marketing, 34*(1), 22–45.

Keegan, B. J., & Rowley, J. (2017). Evaluation and decision making in social media marketing. *Management Decision, 55*(1), 15–31.

Kumar, V., & Mirchandani, R. (2012). Increasing the ROI of social media marketing. *MIT Sloan Management Review, 54*(1), 55.

Kurniawati, K., Shanks, G., & Bekmamedova, N. (2013). The business impact of social media analytics. In *ECIS 2013 Proceedings, AIS Electronic Library* (pp. 1–12).

Lamberton, C., & Stephen, A. (2016). A thematic exploration of digital, social media, and mobile marketing: Research evolution from 2000 to 2015 and an agenda for future inquiry. *Journal of Marketing, 80*(6), 146–172.

Lin, Y. (2020). *10 Twitter statistics every marketer should know in 2020 [Infographic]*. Retrieved 5 June 2020, from https://www.oberlo.com/blog/twitter-statistics.

Makrides, A. (2017). *Assessing the prospects of building brand awareness overseas for a local corporate group via digital marketing* (Unpublished master's thesis). Cyprus International Institute of Management, Limassol, Cyprus.

Makrides, Anna, Vrontis, Demetris, & Christofi, Michael. (2019). The gold rush of digital marketing: Assessing prospects of building brand awareness overseas. *Business Perspectives and Research., 8,* 227853371986001. https://doi.org/10.1177/2278533719860016.

Mediakix. (2019). *Case study: Dunkin' donuts marketing strategy runs on nano-influencers*. Retrieved 18 June 2020, from https://mediakix.com/blog/dunkin-donuts-nano-influencer-marketing-case-study-instagram/.

Modicum. (2016). *Why your brand needs a real style guide*. Retrieved 18 June 2020, from https://www.forbes.com/sites/propointgraphics/2016/07/24/brand-style-guides/#3c15f3f761a5.

Mohanty, P. C., Dash, M., Dash, M., & Das, S. (2019). A study on factors influencing training effectiveness. *Revista Espacios, 40,* 7–15. Retrieved from http://www.revistaespacios.com/a19v40n02/19400207.html.

Mondal, S. R. (2020). A systematic study for digital innovation in management education: An integrated approach towards problem-based learning in Vietnam. In *Digital innovations for customer engagement, management, and organizational improvement* (pp. 104–120). IGI Global. https://doi.org/10.4018/978-1-7998-5171-4.ch006.

Mondal, S., Das, S., Musunuru, K., & Dash, M. (2017). Study on the factors affecting customer purchase activity in retail stores by confirmatory factor analysis. *Revista Espacios, 38*(61).

Mondal, S., Mall, M., Mishra, U. S., & Sahoo, K. (2017). Investigating the factors affecting customer purchase activity in retail stores. *Revista Espacios, 38*(57).

Mondal, S., & Sahoo, K. K. (2020). A study of green building prospects on sustainable management decision making. In *Green building management and smart automation* (pp. 220–234). IGI Global. https://doi.org/10.4018/978-1-5225-9754-4.ch011.

Murdough, C. (2009). Social media measurement: It's not impossible. *Journal of Interactive Advertising, 10,* 94–99.

Nadanyiova, M., & Das, S. (2020). Millennials as a target segment of socially responsible communication within the business strategy. *Littera Scripta, 13*(1), 119–134. https://doi.org/10.36708/Littera_Scripta2020/1/8.

Nair, M. (2011). Understanding and measuring the value of social media. *Journal of Corporate Accounting & Finance, 22*(3), 45–51.

Newberry, C. (2019). *37 Instagram statistics that matter to marketers in 2020.* Retrieved 22 June 2020, from https://blog.hootsuite.com/instagram-statistics/.

Relling, M., Schnittka, O., Sattler, H., & Johnen, M. (2016). Each can help or hurt: Negative and positive word of mouth in social network brand communities. *International Journal of Research in Marketing, 33*(1), 42–58.

Rishi, B., & Bandyopadhyay, S. (Eds.). (2017). *Contemporary issues in social media marketing*. Routledge.

Rosengren, S. (2019). Redefining advertising in a changing media landscape. In M. Deuze & M. Prenger (Eds.), *Making media: Production, practices, and professions* (pp. 389–398). Amsterdam: Amsterdam University Press. https://doi.org/10.2307/j.ctvcj305r.31.

Salesforce. (2018). *State of Marketing Report* (pp. 2–4). Salesforce.

Schmitt, B. (2012). The consumer psychology of brands. *Journal of Consumer Psychology, 22,* 7–17.

Sharma, E., & Das, S. (2020). Measuring impact of Indian ports on environment and effectiveness of remedial measures towards environmental pollution. *International Journal of Environment and Waste Management, 25*(3), 356–380. https://doi.org/10.1504/IJEWM.2019.10021787.

ShortyAwards. (2016). *Dove and Twitter #SpeakBeautiful—The Shorty Awards.* Retrieved 22 June 2020, from https://shortyawards.com/8th/dove-and-twitter-speakbeautiful-2.

Singh, S., & Das, S. (2018). Impact of post-merger and acquisition activities on the financial performance of banks: A study of Indian private sector and public sector banks. *Revista Espacios Magazine, 39*(26), 25.

Singh, L. B., Mondal, S. R., & Das, S. (2020). Human resource practices & their observed significance for Indian SMEs. *Revista ESPACIOS, 41*(7). Retrieved from http://www.revistaespacios.com/a20v41n07/20410715.html.

Singh, S., Mondal, S., Singh, L. B., Sahoo, K. K., & Das, S. (2020b). An empirical evidence study of consumer perception and socioeconomic profiles for digital stores in Vietnam. *Sustainability, 12*(5), 1716.

Singh, I., Nayyar, A., & Das, S. (2019). A study of antecedents of customer loyalty in banking & insurance sector and their impact on business performance. *Revista ESPACIOS, 40*(6).

Singh, I., Nayyar, A., Le, D. H., & Das, S. (2019). A conceptual analysis of internet banking users' perceptions. An Indian perceptive. *Revista ESPACIOS, 40*(14).

Sirgy, J. (1982). Self-concept in consumer behaviour: A critical review. *Journal of Consumer, 9*(3), 287–300.

Smith, P. R., Berry, C., & Pulford, A. (1999). *Strategic marketing communications*. London: Kogan Page.

Spary, S. (2020). Airbnb lets users 'pay it forward' with #OneLessStranger. Retrieved 22 June 2020, from https://www.campaignlive.com/article/airbnb-lets-users-pay-forward-onelessstranger/1327693.

Stephen, A. T., & Galak, J. (2012). The effects of traditional and social earned media on sales: A study of a microlending marketplace. *Journal of Marketing Research, 49*(5), 624–639.

Stewart, M. C., & Arnold, C. L. (2018). Defining social listening: Recognizing an emerging dimension of listening. *International Journal of Listening, 32*(2), 85–100.

Supphellen, M. (2000). Understanding core brand equity: guidelines for in-depth elicitation of brand associations. *International Journal of Market Research, 42*(3), 1–14.

The State of Influencer Marketing 2020: Benchmark Report. (2020). Retrieved 22 June 2020, from https://influencermarketinghub.com/influencer-marketing-benchmark-report-2020/.

Tkaczyk, J. (2016). Digital consumer: Trends and challenges. *IMPACT, 353*.

Trainor, K. J., Andzulis, J. M., Rapp, A., & Agnihotri, R. (2014). Social media technology usage and customer relationship performance: A capabilities-based examination of social CRM. *Journal of Business Research, 67*(6), 1201–1208.

Tuten, T. L., Solomon, M. R., & Rishi, B. (2020). *Social media marketing*. New Delhi: Sage.

US Social Trends for 2020. (2020). Retrieved 3 June 2020, from https://www.emarketer.com/content/us-social-trends-for-2020.

Van, N. T. T., Vrana, V., Duy, N. T., Minh, D. X. H., Dzung, P. T., Mondal, S. R., & Das, S. (2020). The role of human–machine interactive devices for post-COVID-19 innovative sustainable tourism in Ho Chi Minh City, Vietnam. *Sustainability, 12*(22), 9523. https://doi.org/10.3390/su12229523.

Vollero, A., Schultz, D. E., & Siano, A. (2019). IMC in digitally-empowering contexts: The emerging role of negotiated brands. *International Journal of Advertising, 38*(3), 428–449.

Wang, X., Yu, C., & Wei, Y. (2012). Social media peer communication and impacts on purchase intentions: A consumer socialization framework. *Journal of interactive marketing, 26*(4), 198–208.

Yadav, M. (2017). Social media as a marketing tool: Opportunities and challenges. *Indian Journal of Marketing*, *47*(3), 16. https://doi.org/10.17010/ijom/2017/v47/i3/111420.

CHAPTER 11

Sharenting: A New Paradigm of Digital Entertainment of New Age Parenting and Social Media

Ceren Yegen and Subhra R. Mondal

1 Introduction: Sharenting and Its Importance

"Sharenting" is the combination of the English words "share" and "parenting," i.e., "social media parenting." The concept, which characterizes the intensive sharing of children in social media, is highly controversial today (Çimke, Gürkan, & Polat, 2018). Parenting is also very much now a days a social media-dependent activity due to more professionals working full time (Nadanyiova & Das, 2020; Singh, Mondal, Singh, Sahoo, & Das, 2020). Parenting is now a days becomes full-time job (Singh, Mondal, & Das, 2020). So much so that Blogger, YouTuber, and Instagram phenomenon mothers emerging today as a new parent

C. Yegen
Mersin University, Mersin, Turkey
e-mail: cerenyegen@mersin.edu.tr

S. R. Mondal (✉)
Duy Tan University, Da Nang, Vietnam
e-mail: subhraranimondal@duytan.edu.vn

S. Das and S. Gochhait (eds.), *Digital Entertainment*,
https://doi.org/10.1007/978-981-15-9724-4_11

typology with new media, become the subject of current debates about privacy and children's rights. The situation is discussed regarding parents sharing online information, stories, videos, or photos about their children, as well as regarding malicious users. Considering the fact that the information can be accessible in future, which is obviously a cautious condition. However, it is also important that parents watch their children carefully in the context of using social media, which is very popular today.

Social media is acting as a tool for today's digital parenting (Sharma & Das, 2020). According to Eraslan (2019), "We need to watch what our children are doing between 02.00–05.00, which is the criminal time zone. Digital media is very much actively used by adolescent young generation (Das, 2020). Social media is where, parents are keeping an eye for keeping an eye on the young ones (Mondal, 2020). Excessive use of social media is a menace now a days (Mondal & Sahoo, 2020). We should definitely know who our child is talking to on social networks, because the bad guys are cheating our child by pretending they are of the same age." Hereby Eraslan (2019) pointed out another danger in social networks, the danger of being deceived. According to Eraslan (2019), personal information and photographs of children should not be shared, and parents should pattern after the "Japanese family model" where privacy is at the forefront. Because, according to Eraslan (2019), publishing the private opens the way for abuse, with the bulk of child abusers being fed through social media. It is thought that Chinese culture and literature had an important influence on the development of Japanese culture in the "pre-modernization" period (Kıral and Kıral, 2009: 54). For some approaches, the introduction of Chinese culture and civilization to Japan actually started with the introduction of Buddhism to Japan (Örnek, 2015: 488). Japan's development experience after World War II is thought to have inspired the policies of second-generation Asian countries (Yülek, 2014). Located in the east of Asia, between the Pacific Ocean and the Sea of Japan, Japan is a country that should attract Turkey's attention in many contexts. Because both countries are Asian in origin. There is admiration and sympathy in Turkey for the Japanese culture and advanced technology. In the Japanese society, where the family is seen as the basis of both success and failure in social life, family is a phenomenon that shapes and dominates the social life. It is considered that a family model has been adopted in Japan even in management and industrial relations.

Because family logic is very important to them and is still being protected. It is also important that family logic has never been mandated by any institution. Family logic is an expression of preferences of people in Japanese economic organizations. According to some approaches, the family is a social organization, a union, and a community. The family is at the center of the continuation of society with economic solidarity and the social transformation of generations (Bovas, 2003: 65; Kincal, 2016: 2; Kıral and Kıral, 2009: 53; As cited in Aca, 2012: 229; Bhappu, 2001: 162; Gökçe, 1976; Öke- T.C. Başbakanlık Aile Araştırma Kurumu Başkanlığı Aile Politikaları Karşılaştırmalı Ülkeler Panoraması 1991). Therefore, family is very important and private in Japanese society, and there are also measurements and caution about the use of social media by Japanese families (Eraslan, 2019).

Instagram is very much used by teenage young adults now a days (Das & Nayyar, 2020). Digital media is now very much prevalent in all the households with high-speed internet & developed mobile phones (Das, Nayyar, & Singh, 2019). Instagram & Facebook are part of everyone's life (Das & Nayyar, 2019). The Web that came into our lives in the early 2000s has become much more personal today (Abdulahi, Samadi, & Gharleghi, 2014: 133) along with Social Network Sites (SNS) adopted by the masses, ultimately inviting a controversial concept such as "*sharenting*." Considering that a Pew Internet survey found that 79% of American adults use the internet and related research found that Facebook dominates the SNS domain (Hampton, Goulet, Rainie, L., & Purcell, 2011), it is of significant importance for almost every segment of society to think about practices in social networks. Because there is also a big problem with social networks, that is, people share too much information. If children give too much information about themselves and their personal lives, it means they can become partially vulnerable and it is necessary to tell them (McAfee, social Networking sites which act as parents' guide). Therefore, in this study, the concept of *sharenting*, refers to *sharing parenting*, will be discussed and the relationship between social media and privacy will be addressed. In this way, it will be possible to express an up-to-date perspective on the issue and to reveal possible risks. The study will also examine shares made under the hashtag #sharenting on Instagram, in terms of privacy and parenting. The analysis method of this study is content analysis.

2 Social Networking Services—Social Media

The speed of the internet, spread rapidly around the world and has deeply affected many fields since its invention (Sözbilir & Dursun, 2018: 186), and changed the way people receive information, the excitement of multimedia has made the internet hugely popular. In today's world, the internet has an important place in human lives for which it is hard to imagine someone who doesn't check for updates on social networks at least once a day. Because in reality we are living, and requires staying in touch with the latest news and trends (Talaue, AlSaad, AlRushaidan, AlHugail, & AlFahhad, 2018: 27; Niedzviecki, 2019). Perceptions of parents are changing rapidly too (Singh, Nayyar, Le, & Das, 2019). In every aspect of life digital media has a long term effect (Mohanty, Dash, Dash, & Das, 2019).

Social networking sites are the web services that allow people to create a public or semi-public profile in a finite system, with users whom they share connections and views. (Abdulahi et al., 2014: 133). While Instagram is rapidly entering into every young adult's daily life, other similar apps are slowly innovating themselves (Singh, Nayyar, & Das, 2019). All the households in world are under the influence of social media & entertainment (Gupta, Jena, Samantaray, & Das, 2019). Everyone in all household is affected by constant media exposure (Singh & Das, 2018). Emotional stability is very much affected by constant influence of digital media (Jain, Jain, & Das, 2018).

Social networking services are the websites that provide a virtual community for people, interested in a particular topic or wants to hang out with each other. Children and young people love social networking sites because they use these platforms to chat, to send emails, and to share photos with friends and family online. Social networking sites are actually new media platforms where interaction is essential, and they are quite popular today. According to a previously noted Pew survey, 52% of Facebook users and 33% of Twitter users interact with media platform, while only 7% of MySpace users and 6% of LinkedIn users do the same (Hampton, Goulet, Rainie, & Purcell, 2011; A parent's Guide to Social Networking Sites, McAfee). Social media has a structure, in which social communication is provided through words, images, and audio files, and individuals share their stories and experiences in certain patterns within the framework of this structure (Menteşe, 2013: 2). According to Niedzviecki (2019) in fact, social networking services are not very different from the platforms that provide individuals to meet and flirt over the internet.

Parenting is now becoming a digitalized supported job with new apps & innovations (Das, Mondal, Sahoo, Nayyar, & Musunuru, 2018). Instagram, Snapchat, Twitter, & TikTok are now a days becoming a rage in young adults (Mondal, Mall, Mishra, & Sahoo, 2017). The psychological makeup of young adults is very much dynamic & affected by a lot of media exposure (Mondal et al., 2017). So, digitally all the families are now affected (Das & Mondal, 2016). Social media is a concept that is influential on direct participation of the individual in the communication process and stands for the internet-based information dissemination/sharing mediums such as Facebook, Twitter, Instagram, YouTube (Uluç & Yarcı, 2017: 88–89). Social media is a phenomenon that transforms the interaction and communication of individuals in the world, and some approaches argue that social media as not a new concept. According to these approaches, sociality has been present since the first interaction of human beings. As per the social media, it has been a concept which has affected many aspects of human communication, becoming one of the indispensable applications of everyday life. Major social networking sites such as Facebook, YouTube, Twitter, My Space, CyWorld, and Lunar Storm that emerged in the twenty-first century actually represent the current state of individual-based interaction (Edosomwan, Kalangot Prakasan, Kouame, Watson, & Seymour, 2011).

Social media also qualifies as a popular and influential space for advertising and marketing today (Siddiqui & Singh, 2016: 72). According to Türk (2014: 63), before social media was invented, the fact that an event occurring anywhere could have repercussions worldwide was proportional to the fact that the traditional media included the event and kept it on the agenda. But in parallel with the developments in the new communication technologies today, individuals now have the opportunity to set their own mass media and agendas. Today, the use of social media has exploded globally (Otero, 2017: 412), and new technologies have also helped the development of certain phenomena such as democratic participation, e-learning, and e-shopping (Zeitel-Bank & Tat, 2014: 1184).

Social media is a habit for many of us. It is an environment that mediates socialization, also refers to a consumption platform. So much so that individual and corporate sales are made on social media (especially Instagram) today. Thus, studies dealing with social networks as a consumption and shopping medium are also carried out. These studies can also deal with facts such as hedonism, consumer behavior models, and consumption habits. This situation shows us how consumption habits have changed and transformed through new media. With new media environments/social networks, individuals and many habits are changing and shaping.

Social media, can be described as a means of social communication (Bowie, 2012), but also an area in which we socialize. Before the social media, we were speaking about traditional media, one-directional flow of information and communication, and public media (How the World Changed Social Media, 2016), but now we can even talk about "public opinion publication" through social networks. So much so that, using the social media environments and applications, individuals in the public can make their own publications/records, share them with masses and even become phenomena. This is certainly remarkable not only for the interaction-based structure of social media but also for the traditional media as well as the ordinary users.

It is obvious that social networks affect and shape our lives. So much so that individuals from almost every age group use new media and include them in their daily life. For this reason, there are many valuable studies on social media's entering people's daily lives and their effects. Carrying out these studies is important in terms of providing current data on the adaptation of the society to social media. However, it is important and necessary to use social media correctly and beneficially.

We also know that social media mediate various virtual identity formations. Individuals, especially young people, can engage themselves in any identity they want through social networks. Users can even use fake or anonymous identities. Identity and self are transforming through social networks. Ideal identities are designed through social networks. The user/individual determines the limits of this design. However, the sense of belonging can also be shaped in online conditions (Güz & Yegen, 2017; Yegen & Kırık, 2019).

3 Risks in Social Networking

Social networks are one of the current grounds of many essential practices, from sports to politics. Today, many youngsters spend most of their time in digital media, as they follow the developments in mobile communication technologies and the internet as "young consumers" (Çakır, Çakır, & Eru, 2013: 88).

According to Karahisar (2015: 271), in our day of extreme exhibitionism, protecting personal data becomes more difficult in each passing day. According to him, as Poster says, the exhibitionism is voluntary today, and individuals do not hesitate to post their confidential and private environments on social media, on the contrary, they enjoy doing it. However, sharing information or photos/videos on social networks can lead to problems on privacy and confidentiality. Additionally, it is important to note that there are certain cases of theft due to social networks and applications that often allow location/position reporting. Furthermore, social media fraud is also evident today as a digital fraud.

The risks of social networks are not for in the individuals themselves, but also for the family institution, and the children who are the favorites of that institution. Today, experts in many areas warn parents to keep track of their children's social media account, use and also to take care of their children's privacy in using it.

It may also be suggested that the following measures should be taken against the risks for children, particularly in the social networks:

- Set limits and rules for online behavior.
- Limit the amount of time that children spend on the internet.
- Discuss what is appropriate and what is not for online sharing.
- Advise your child to be aware of anyone they don't know and want to join their network.
- Tell your child not to meet face-to-face with people he/she meets online.
- Insist your child tell you when he/she notices something strange or unusual, feels uncomfortable or threatened.
- Teach your child to be wary of emails that contain questions, offers or links to websites (A Parent's Guide to Social Networking Sites, McAfee.).

It should also be added that, due to the social networking sites, certain problems at different types and frequencies may arise in preteens (9–12) and teens (13–15) children such as individual safety, identity stealing which are disturbing and an increase in mental health issues like depression, social isolation, anxiety, and poor sleep. There is also a risk for children to encounter cyberbullying or to become sex targets. Some of these risks are a result of less guidance and parenting with a lack of proper knowledge about online safety procedures. Additionally, social media can also ensure children to interact with both their daily social sphere and with certain groups such as (LGBTI + etc.), building stronger friendships. It also improves access to diversity, increases incorporation and reduces social isolation in our postmodern age. To take positive advantage of technology, children, along with their parents, need to be aware of these possible risks of social media and how to drive them (Dyer, 2018). Providing attractive communication tools for the daily life, e-learning mediums and social networking technologies allows students to create a digital identity and to express themselves digitally; in addition, it mediates students to be socially interactive and to create an online learning community (Bozkurt, Karadeniz, & Kocdar, 2017: 348). It seems that e-learning environments will be discussed further as information and communication technologies progress (Das, 2021; Duy et al., 2020; Van et al., 2020).

4 A New Parenting Typology: "Sharenting" and Privacy Discussions

"Sharenting" is the tendency of parents to share pictures of their children on social media sites. *Sharenting* has been described by the Collins Dictionary as "the habitual use of social media to share detailed information of one's children" (Brosch, 2016: 225–226). This increase the controversy, as parents have to balance their right to share and children's right to privacy (Wagner & Gasche, 2018: 977). A study based on interviews with 17 parent bloggers reveals that bloggers deal with deep ethical dilemmas. These individuals, who determine their children's identities in digital terms, constitute the public aspects of their children's lives (Blum-Ross & Livingstone, 2017). The terms "over-sharing" and "parenting" were first used by Steven Leckart, who wrote in The Wall Street Journal in 2012. *Sharenting* is also used in literature as "oversharenting," extreme social media parenting. The counterpart of this concept in Turkish is expressed as "over-sharing and parenthood" (Maraşlı, Suhendan, Yılmaztürk, & Cok, 2016: 399; Ozan Leymun and Kurt, 2018).

Considering that the parents of our age, who grow up their children in the first period of the digital culture and who face new parental problems compared to previous generations, are the role models for their children in terms of using social media, (Ante-Contreras, 2016), new concerns arise about being the right role model for the children. As in the next-generation parenting becomes popular, parents will share the joy and challenges of parenting on social networks, documenting the lives of their children in public with a frequency that has become the norm (Brosch, 2016: 225–226).

According to some approaches, individuals are making sharings for the following reasons:

- Linking
- Self-expression
- Getting advice/giving advice
- Sharing all kinds of experiences
- Raising awareness/attracting attention or raising funding
- Virtual communities, reaching out to others in the groups (Greaves, 2018).

A recent study suggests that *sharenting* may be related to narcissism, even jealousy. Similarly, a result from the survey implemented in this study demonstrated that 3% of mothers shared photos of their children because they intended to be coveted by others. This result, which sounds surprising at first vison, was found by researchers to be explained by Foster's theory of self-confidence. Because Foster pointed out that indi viduals like to perform better than others, and considered that individuals value their success or superiority over others (Wagner & Gasche, 2018: 985). It is also important to note that though sharing on social networks is only shared with a group of friends, it should also be considered that not everyone on the list of friends is probably well known (Çimke et al., 2018: 265).

While discussions were ongoing about whether "Sharenting" would lead to a violation of private information that could put children at risk (C.S. Mott Children's Hospital, 2015), another study was conducted in 2016 about Facebook usages of 168 Polish parents demonstrated that *sharenting* was a common exercise among them (Brosch, 2016). A study by Maraşlı, Suhendan, Yılmaztürk, and Cok (2016) also supported the existence of this new trend among Turkish. The findings revealed that some parents shared their children's detailed experiences on social media (Maraşlı et al., 2016: 405). As can be seen from many phenomena, the social network usages of parents can also be related to advertising, product guidelines for pregnancy and parenting (Lupton, Pedersen, & Thomas, 2016). It is so apparent today that many children are acquiring a digital

ID before they even talk, and the parents of our day are not only creating digital lives to their children through Facebook, Instagram, or blogs, but also building digital and online lives for them (Damkjaer, 2018).

Young children may be affected by their parents' views on use of social media, as they may not be aware of how its use affects their children (Life in "likes" Children's Commissioner report into social media use among 8–12 year-olds). When parents share their children information online, without their consent (Steinber, 2017). Although families do *sharenting* in good faith, involuntarily created digital footprints can create a number of rights violations (Tuğtekin & Odabaşı, 2017). The "digital footprint" characterizes all traces transferred to electronic media by pressing, clicking or touching the keyboard, mouse, or screen. In fact, almost everything about the user on the internet is their digital footprints. Digital footprint can be used, without the consent of the user, to extract information such as personal information, religious views, political views, and family and relationships, exposing people's privacy to the public sphere (Dönmez, 2019). Parents can unwillingly violate their children's right to privacy, forgotten, and even play a mediating role in exploiting it by other people (Çimke et al., 2018: 261).

Children growing up within *sharenting* environments today will surely be influenced by public interactions that depict curatorship about their identity. Oversharing of individual childhood moments creates a moderated digital identity in the child. *Sharenting* (Davidson-Wall, 2018), the new vision in parenting, can cause parents to be unaware of hidden dangers, even if they believe they are keeping their children safe (Dyer, 2018). Today, Facebook, Twitter, YouTube as well as Instagram are popular *sharenting* mediums. Following the *sharenting* trend, parents, especially social media phenomenon or influencer, blogger/YouTuber parents, share almost every moment of their children on social networks in Turkey, as in many parts of the world. Undoubtedly, many children, who grow up today, will have the chance to see how their lives evolve through social media. However, the feeling of the children is still unknown when they see their personal lives is done online by their parents (Moser, Chen, & Schoenebeck, 2017). Families should therefore be thoroughly informed about *sharenting*, what is not shared in general should not be shared online (Otero, 2017: 412).

5 Research Methodology

In this part of the study, shares including the hashtag #sharenting, available on Instagram as of March 31, 2019, were examined concerning privacy and parenting. A total of 4720 shares were found in a scan on Instagram on the relevant date with the hashtag #sharenting. Since examining all of these shares would not limit the research, a content analysis was conducted for the last 100 shares which were selected through the simple random sampling method. Content analysis enables systematic and qualitative analysis of the data (Bilgin, 2006; Gökçe, 2001).

The analysis showed that only one quarter (25) of the shares (100) were photographed by children alone. Photos of children shared with a female or a mother figure, which may be considered to be parents, are also common Table 1.

In the 20 out of the 25 only-child photos, the faces of the children were visible, while they were blurred in five photos as per Table 2.

There are children in 68 out of 100 shares examined in the study. The photos of the children were generally shared together with females and males who can be considered as their parents Table 3.

In the majority of the 43 photos shared including Female & Male & Child Together, the emotional states of the subjects are unclear Table 4.

Table 1 Shares on Instagram with the Hashtag #sharenting

Share type	*Frequency*
The Photo of Only a Child	25
Female (Mother etc.) & Child Photo	26
Male (Father etc.) & Child Photo	6
The Photo of Female & Male & Child Together	11
Writing	16
Visual	10
Video	6
Total	**100**

Table 2 Types of the child photos

Child's face is visible	*Child's face is blurred*
20	5

Table 3 Subjects of the share

Only-child	*Female & male & child together*
25	43

Table 4 Apparent emotional state of the subjects

Happy	*Unhappy*	*Unclear emotional state*
18	1	24

Table 5 Apparent emotional states of children

Happy	*Unhappy*	*Unclear emotional status*
11	4	10

Table 6 Photo environments

Indoors	*Outdoors*
14	29

In the 25 only-child photos of children, the emotional states of the children/babies are mostly in the category of happy. The children laughed and smiled at most of the photos that were shared. However, it was observed that emotional state of the child was uncertain in 10 shares Table 5.

The majority of the 43 photos including Female & Male & Child Together were taken outdoors; 29 were taken outdoors while 14 of them were taken indoors Table 6.

6 Conclusion and Suggestions

It is a common opinion that social media is getting more in demand gradually. It should be seen that new practices that are popular in daily life are of great importance through social networks. So much so that social life is shaped by technology and technology-mediated behaviors and practices. The concept of "sharenting" is also a very important and popular trend, even a matter of concern. Because, many moments and details regarding the lives of children and babies in need of protection in social networks can be shared by their families and invitations to possible risks. Perhaps the most important measure in terms of risks that come with Sharenting is to think in detail about what to share and not to share without forgetting that social networks are public.

In this study, it is important that a total of 4720 sharings were found in the scan with the hashtag #sharenting on Instagram, which is one of the well-liked social networks of today. A lot of sharing points increase interest in this trend. While hashtags which created by Twitter users in 2007 and first used by Chris Messina, characterize clickable tags on social networks, are the efficient way to investigate a topic, it also describes a digital narrative tool in itself. It is a functional tool of digital marketing (Adres Gezgini, 2017, 27 March; Soydan, 2015; Yegen, 2018). Therefore, in this study, it was possible to reach a large number of sharings by using the hashtag #sharenting.

In this study, it was observed that in the shares including the hashtag #sharenting on Instagram, the faces of children were generally visible, and the photos of children together with males and females, possibly their parents, were generally posted. This may suggest that surveillance is prioritized concerning *sharenting*. On December 5, 2019, it was stated that Instagram officially banned users under 13 from using the app. The study found that in the majority of the shares, outdoors were preferred for shooting the photo, and children were photographed with females (possibly a mother, caregiver, or family individual) more than males. Since the emotional states of the subjects in the shares appear to be happy in general, it might suggest that the sharing is enjoyed.

As a result, it can be said that sharenting is popular and current on Instagram. However, with the further popularization of sharenting in the future, it can be thought that the number of researches on this subject will increase, as will the concept-related posts. At this point, the role of concept photography types such as birthing photography and birthday photography should be considered on the popularization of sharenting.

References

"An Instagram, 13 Yaşından Küçüklerin Platformu Kullanmasını Yasakladı". Berk Baylançiçek (2019, December 5). https://www.webtekno.com/instagram-13-yas-siniri-h81126.html.

A publication from C.S. Mott Children's Hospital, the University of Michigan Department of Pediatrics and Communicable Diseases, and the University of Michigan Child Health Evaluation and Research (CHEAR) Unit, Volume 23, Issue 2, March 16, 2015.

Abdulahi, A., Samadi, B., & Gharleghi, B. (2014). A study on the negative effects of social networking sites such as Facebook among Asia Pacific University Scholars in Malaysia. *International Journal of Business and Social Science, 5*(10), 133–145.

Aca, Z. (2012). Düşey Toplum Kuramı Yaklaşımıyla Japon Çalışma Kültürünü Anlamak. *Organizasyon Ve Yönetim Bilimleri, 4*(2), 225–234.

Adres Gezgini. (2017, March 27). "Sosyal Medyada Hashtag Kullanımı". https://blog.adresgezgini.com/sosyal-medyada-hashtag-kullanimi.

Ante-Contreras, D. (2016). *Distracted parenting: How social media affects parent-child attachment* (Theses, Projects, and Dissertations, Paper 292).

A Parent's Guide to Social Networking Sites, McAfee, Erişim. (2019, March 22). https://www.toshiba.eu/mcafee_safesocialnetworks/socialnetworking_EN.pdf.

Bhappu, A. D. (2001). Japon Ailesi: Japon Şirket Ağları ve Japon Yönetimi İçin Kurumsal Bir Mantık. *Akdeniz İ.İ.B.F. Dergisi*, 2, 156–167.

Bilgin, N. (2006). *Sosyal Bilimlerde İçerik Analizi Teknikler ve Örnek Çalışmalar*. Ankara: Siyasal Bookstore.

Blum-Ross, A., & Livingstone, S. (2017). Sharenting: parent blogging and the boundaries of the digital self. *Popular Communication, 15*(2), 110–125.

Bowie, S. (2012, July 8–12). Social work and the role of social media best practices. Presentation: Association of Caribbean Social Work Educators (ACSWE) 11th Biennial Conference Curacao.

Bozkurt, A., Karadeniz, A. & Kocdar, S. (2017). Social networking sites as communication, interaction, and learning environments: Perceptions and preferences of distance education students. *Journal of Learning for Development-JL4D, 4*(3), 348–365.

Brosch, A. (2016). When the child is born into the Internet: Sharenting as a growing trend among parents on Facebook. *The New Educational Review, 43*, 225–235.

Büyükbaş, H. (2003). Japon Modernleşmesi Üzerine (1868–1912). *Bilimname, III*(3), 65–86.

Çakır, F., Çakır, M., & Eru, O. (2013). The effect of the social media on young consumers' consumption expenditure. *International Journal of Social Sciences and Humanity Studies, 5*(2), 86–96.

Çimke, S., Gürkan, Y. D., & Polat, S. (2018). Sosyal Medyada Çocuk Hakkı İhlali: Sharenting. *Güncel Pediatri JCP, 16*(2), 261–267.

Damkjaer, M. S. (2018). Sharenting = Good parenting? Four parental approaches to sharenting on Facebook. In G. Mascheroni, C. Ponte & A.

Jorge (Eds.), *Digital parenting, the challenges for families in the digital age* (pp. 209–218). Göteborg: Nordicom.

Das, S. (2020). Innovations in digital banking service brand equity and Millennial consumerism. In *Digital transformation and innovative services for business and learning* (pp. 62–79). IGI Global. https://doi.org/10.4018/978-1-7998-5175-2.ch004.

Das, S. (2021). *Search engine optimization and marketing: A recipe for success in digital marketing* (1st ed.). CRC Press. https://doi.org/10.1201/9780429298509.

Das, S., & Mondal, S. (2016). *Yours digitally: Patanjali*. Available at SSRN 2737299.

Das, S., Mondal, S. R., Sahoo, K. K., Nayyar, A., & Musunuru, K. (2018). Study on impact of socioeconomic make up of Facebook users on purchasing behavior. *Revista Espacios, 39,* 28–42. Retrieved from http://www.revistaespacios.com/a18v39n33/18393328.html.

Das, S., & Nayyar, A. (2019). Innovative ideas to manage urban traffic congestion in cognitive cities. In *Driving the development, management, and sustainability of cognitive cities* (pp. 139–162). IGI Global. https://doi.org/10.4018/978-1-5225-8085-0.ch006.

Das, S., & Nayyar, A. (2020). Effect of consumer green behavior perspective on green unwavering across various retail configurations. In *Green marketing as a positive driver toward business sustainability* (pp. 96–124). IGI Global. https://doi.org/10.4018/978-1-5225-9558-8.ch005.

Das, S., Nayyar, A., & Singh, I. (2019). An assessment of forerunners for customer loyalty in the selected financial sector by SEM approach toward their effect on business. *Data Technologies and Applications, 53*(4), 546–561. https://doi.org/10.1108/DTA-04-2019-0059.

Davidson-Wall, N. (2018). *"Mum, seriously!": Sharenting the new social trend with no opt-out*. Debating Communities and Social Networks 2018 OUA Conference.

Dönmez, A. (2019, January 25). "DİJİTAL AYAK İZİ". https://prezi.com/8xkqzcx5oldv/dijital-ayak-izi/.

Duy, N. T., Mondal, S. R., Van, N. T. T., Dzung, P. T., Minh, D. X. H., & Das, S. (2020). A study on the role of web 4.0 and 5.0 in the sustainable tourism ecosystem of Ho Chi Minh City, Vietnam. *Sustainability, 12*(17), 7140. https://doi.org/10.3390/su12177140.

Dyer, T. (2018). *The effects of social media on children*. djim.management.dal.ca | Vol. 14.

Edosomwan, S., Kalangot Prakasan, S., Kouame, D., Watson, J., & Seymour, T. (2011). The history of social media and its impact on business. *The Journal of Applied Management and Entrepreneurship, 16*(3), 79–91.

Erarslan, L. (2019, March 17). "Ailelere sanal alemde Japon aile modelini öneriyorum". https://www.risalehaber.com/ailelere-sanal-alemde-japon-aile-modelini-oneriyorum-349400h.htm.

Gökçe, B. (1976). Aile ve Aile Tipleri Üzerine Bir İnceleme. *Hacettepe Sosyal ve Beşeri Bilimler*, *8*(1–2).

Gökçe, O. (2001). *İçerik Çözümlemesi Teori-Metod-Uygulama* (Extended 3rd ed.). Konya: Nurol Printing.

Greaves, D. (2018). *Supporting parents make better decisions if sharing online about their children: "Respectful sharenting"*. NCP Conference.

Gupta, D. K., Jena, D., Samantaray, A. K., & Das, S. (2019). HRD climate in selected public sector banks in India. *Revista ESPACIOS*, *40*(11).

Güz, N., & Yegen, C. (2017). Ölü Canlar'dan Instagram'a Kimlik ve Statü Sahibi Olmanın Allomorfizmi Üzerine. *İletişim Kuram Ve Araştırma, 45,* 103–123.

Hampton, K.N., Goulet, L. S., Rainie, L., & Purcell, K. (2011). *Social networking sites and our lives How people's trust, personal relationships, and civic and political involvement are connected to their use of social networking sites and other Technologies*. Pew Research Center's Internet & American Life Project.

How the World Changed Social Media. (2016). Daniel Miller Elisabetta Costa Nell Haynes Tom McDonald Razvan Nicolescu Jolynna Sinanan Juliano Spyer Shriram Venkatraman Xinyuan Wang, First published in 2016 by UCL Press University College London Gower Street London WC1E 6BT.

Jain, S., Jain V., & Das, S. (2018). Relationship analysis between emotional intelligence and service quality with special evidences from Indian banking sector. *Revista ESPACIOS*, *39*(33).

Karahisar, T. (2015). Yeni Medya, Mahremiyet ve Unutulma Hakkı. In *AĞDAKİ ŞÜPHE Bir Sosyal Medya Eleştirisi* (T. Kara & E. Özgen, Compl., pp. 249–276). İstanbul: Beta.

Kıncal, A. (2016). Japon Kalkınması: Tarihsel Süreç ve Politikalar. *DERIN Dokuz Eylül Üniversitesi Araştırma ve İnovasyon, 6,* 76–86.

Kıral, B., & Kıral, E. (2009). Japonya İlköğretim Sistemi ve Türkiye İlköğretim Sisteminin Karşılaştırılması. *Journal of Dicle University Ziya Gökalp Faculty of Education, 12,* 53–65.

Life in 'Likes': Children's Commissioner Report into Social Media Use Among 8–12 Year Olds, Erişim. (2019, March 28). https://www.childrenscommissioner.gov.uk/wp-content/uploads/2018/01/Childrens-Commissioner-for-England-Life-in-Likes.pdf.

Lupton, D., Pedersen, S. & Thomas, G. M. (2016). Parenting and digital media: From the early web to contemporary digital society. *Sociology Compass, 10/8,* 730–743.

Maraşlı, M., Suhendan, E., Yılmaztürk, N. H., & Cok, F. (2016). Parents' shares on social networking sites about their children: Sharenting. *Anthropologist, 24*(2), 399–406.

Menteşe, M. (2013). *Sosyal Medya Ortam ve Araçlarının Eğitimde Kullanımına İlişkin Okul Yöneticilerinin ve Öğretmenlerin Görüşleri* (Master's thesis). Hacettepe University, Institute of Social Sciences, Educational Sciences, Educational Administration, Inspection, Planning and Economics, Ankara.

Mohanty, P. C., Dash, M., Dash, M., & Das, S. (2019). A study on factors influencing training effectiveness. *Revista Espacios, 40,* 7–15. Retrieved from http://www.revistaespacios.com/a19v40n02/19400207.html.

Mondal, S. R. (2020). A systematic study for digital innovation in management education: An integrated approach towards problem-based learning in Vietnam. In *Digital innovations for customer engagement, management, and organizational Improvement* (pp. 104–120). IGI Global. https://doi.org/10.4018/978-1-7998-5171-4.ch006.

Mondal, S., Das, S., Musunuru, K., & Dash, M. (2017). Study on the factors affecting customer purchase activity in retail stores by confirmatory factor analysis. *Revista Espacios*, *38*(61).

Mondal, S., Mall, M., Mishra, U. S., & Sahoo, K. (2017). Investigating the factors affecting customer purchase activity in retail stores. *Revista Espacios*, *38*(57).

Mondal, S., & Sahoo, K. K. (2020). A study of green building prospects on sustainable management decision making. In *Green building management and smart automation* (pp. 220–234). IGI Global. https://doi.org/10.4018/978-1-5225-9754-4.ch011.

Moser, C., Chen, T., & Schoenebeck, S. (2017, May, 6–11). *Parents' and children's preferences about parents sharing about children on social media.* CHI 2017. Denver.

Nadanyiova, M., & Das, S. (2020). Millennials as a target segment of socially responsible communication within the business strategy. *Littera Scripta, 13*(1), 119–134. https://doi.org/10.36708/Littera_Scripta2020/1/8.

Niedzviecki, H. (2019). *Dikizleme Günlüğü* (2nd ed., G. Gündüç, Trans.). İstanbul: Ayrıntı Publications.

Örnek, S. V. (2015). Japonya ve Türkiye'deki, Dini, Kültürel ve Sosyal Reformlara Bir Bakış. *Folklor/Edebiyat, 21*(82), 487–501.

Otero P. (2017). Sharenting... should children's lives be disclosed on social media? *Arch Argent Pediatr, 115*(5), 412–413.

Ozan Leymun, Ş., & Kurt, A. A. (2018, April 11–13). *A new syndrome in families: Sharenting*. International Child and Information Safety Congress "Digital Games". Ankara.

Sharma, E., & Das, S. (2020). Measuring impact of Indian ports on environment and effectiveness of remedial measures towards environmental pollution.

International Journal Environment and Waste Management, 25(3), 356–380. https://doi.org/10.1504/IJEWM.2019.10021787.

Siddiqui, S., & Singh, T. (2016). Social media its impact with positive and negative aspects. *International Journal of Computer Applications Technology and Research, 5*(2), 71–75.

Singh, S., & Das, S. (2018). Impact of post-merger and acquisition activities on the financial performance of banks: A study of Indian private sector and public sector banks. *Revista Espacios Magazine, 39*(26), 25.

Singh, L. B., Mondal, S. R., & Das, S. (2020). Human resource practices & their observed significance for Indian SMEs. *Revista ESPACIOS, 41*(7). Retrieved from http://www.revistaespacios.com/a20v41n07/20410715.html.

Singh, S., Mondal, S., Singh, L. B., Sahoo, K. K., & Das, S. (2020b). An empirical evidence study of consumer perception and socioeconomic profiles for digital stores in Vietnam. *Sustainability, 12*(5), 1716.

Singh, I., Nayyar, A., & Das, S. (2019). A study of antecedents of customer loyalty in banking & insurance sector and their impact on business performance. *Revista ESPACIOS, 40*(6).

Singh, I., Nayyar, A., Le, D. H., & Das, S. (2019). A conceptual analysis of internet banking users' perceptions. An Indian perceptive. *Revista ESPACIOS, 40*(14).

Soydan, B. (2015, April 28). *Markaların # stratejileri veya hashtag pazarlaması.* http://www.turkishtimedergi.com/dijital-ekonomi/markalarin-stratejileri-veya-hashtag-pazarlamasi/.

Sözbilir, F., & Dursun, M. K. (2018). Does social media usage threaten future human resources by causing smartphone addiction? A study on Students Aged 9-12. *Addicta: The Turkish Journal on Addiction*, 5, 185-203.

Steinber, S. B. (2017). Sharenting: Children's privacy in the age of social media. *Emory Law Journal, 66,* 839–884.

Talaue, G. M., AlSaad, A., AlRushaidan, N., AlHugail, A., & AlFahhad, S. (2018). The impact of social media on academic performance of selected college students. *International Journal of Advanced Information Technology (IJAIT), 8*(4/5), 27–35.

T.C. Başbakanlık Aile Araştırma Kurumu Başkanlığı Aile Politikaları Karşılaştırmalı Ülkeler Panoraması. (1991). Bilim Serisi, Genel Yayın No: 70, Dizgi-Baskı Türk Tarih Kurumu Basımevi, Ankara.

Tuğtekin, U., & Odabaşı, H. F. (2017). 'Sosyal Medya Aileleri'nin Yeni Sendromu: 'Sharenting'. 11th International Computer Instructional Technologies Symposium.

Türk, M. S. (2014). Sosyal Medya ve Gerçeklik. *Kamuda Sosyal Politika* (pp. 63–68), Year 8, No. 17, April–May, June 2014.

Uluç, G., & Yarcı, A. (2017). Sosyal Medya Kültürü. *Dumlupınar University Journal of Social Sciences, 52,* 88–102.

Van, N. T. T., Vrana, V., Duy, N. T., Minh, D. X. H., Dzung, P. T., Mondal, S. R., & Das, S. (2020). The role of human–machine interactive devices for post-COVID-19 innovative sustainable tourism in Ho Chi Minh City, Vietnam. *Sustainability, 12*(22), 9523. https://doi.org/10.3390/su12229523.

Wagner, A., & Gasche, L. A. (2018, March 6–9). *Sharenting: Making decisions about other's privacy on social networking sites*. Multikonferenz Wirtschaftsinformatik (pp. 977–988). Lüneburg, Germany.

Yegen, C. (2018). American Fundamentalism in the new media: Transmedia narratives of Baptists, Evangelists, and Methodists. In R. Yilmaz, M. Nur Erdem F. Resuloglu (Eds.), *Handbook of research on transmedia storytelling and narrative strategies* (pp. 471–486). Philadelphia, PA: IGI Global.

Yegen, C., & Kırık, A. M. (2019). *Sosyal Ağlar ve Kimlik: İdeal "Ben"*. First International Şişli Science Congress, 2019-10-24, 2019-10-25, İstanbul, Turkey.

Yülek, M. (2014). Kalkınma Tartışmalarında Japon Modelinin Yükselişi, Düşüşü ve Tekrar Yükselişi. In K. A. Akkemik, & S. Ünay (Eds.), *Doğu Asya Ekonomi Politiği: Kalkınma, Siyaset, Jeostrateji*. İstanbul: Boğaziçi University.

Zeitel-Bank, N., & Tat, U. (2014, June 25–27). *Social media and its effects on individuals and social systems*. Human Capital without Borders: Knowledge and Learning for Quality of Life. Portorož, Slovenia.

CHAPTER 12

Building Effective Social Media Strategy: Case-Based Learning and Recommendations

Kumar Shalender

1 Introduction

In this era of digital disruption, it is quite a common scenario to find organizations going full steam on their social media initiatives. The popularity of social media has witnessed exponential growth over the past few years and its appeal among young consumers is specifically significant. No wonder, companies across the business segments are using social media platforms to add new customers to their kitty while retaining their existing lot. We interviewed scores of senior-level brand executives across business categories and from our interaction, we find that almost all are unanimous on the need to integrate social media platforms in their branding strategy. However, it occurs to us that many of these branding people lack in clarity of what social media integration actually means. To put forth the dilemma, a brand manager of an automobile company confided in us that despite initiating a dozen of social media campaigns, he and his department are totally clueless what the social media integration means

K. Shalender (✉)
Chitkara Business School, Chitkara University, Punjab, India
e-mail: kumar.shalender@chitkara.edu.in

S. Das and S. Gochhait (eds.), *Digital Entertainment*,
https://doi.org/10.1007/978-981-15-9724-4_12

for prospective customers, how it will affect firm's performance, and more importantly, whether these campaigns will be able to effectively gel with the rest of promotional strategies of the organization or not. Another Brand head of a consumer durable company hesitatingly says that we are promoting big time on social media but aren't able to synergize our branding campaigns on electronic and print media with our social media push.

The reasons behind these dilemmas are not difficult to understand. Social media has been around for quite a while but the breakneck speed of the digital innovations in the last half-a-decade has caught most of the organizations by surprise. Yes, everyone is aware of the importance of social media but the real difficulty arises when one asks how to achieve its seamless integration with other elements of IMC. Today smart devices can easily fit in your pocket and social media applications that run on these devices are accessible 24X7.

Further, sensors embedded on these devices facilitate real time data exchange, thereby allowing users to filter irrelevant information and messages. This enhanced control and awareness is a challenge for branding strategists who are already struggling to cope with new innovations, decreasing loyalty, increasing competition, and tightening regulations. In such an environment, it's absolutely mandatory for organizations to integrate their social media strategy with the rest of the branding techniques in a manner that helps the company to attract new customers while keeping the old ones happy and satisfied.

In the following section, I present guidelines for the seamless integration of social media with the rest of the elements of the IMC strategy. These recommendations are based on interviews of 48 senior-level brand executives and a slew of corporate cases from different industries. Our interactions with these top echelons help bust myths and guide brand managers to achieve desired results with their social media integration strategy.

2 Integrate Social Media to Serve Your Customers Better

Successful organizations deeply know their customers and use social media feedback to serve their customers better. As one of the senior executives of the mobile handset maker told us, "The integration of social media has brought some important inputs, especially from the design viewpoint. We are in the mobile industry for around 20 years now but rapidly evolving rules of customer engagements have prompted us to reimagine our IMC strategy with special focus on social media. We haven't changed dramatically on our products/services but thanks to customers' responses on our social media platforms; we have been able to make small tweaks in our entire value chain which has started yielding rich dividends for us now."

This is exactly what online apparel retailer Myntra did with its social media strategy. The Walmart-owned brand has been able to capture the attention of customers by providing them expert advice on styling trends, lifestyle issues, dietary recommendations, pet care, and safety tips. These smartly conceived conversations married with witty content help the company to gain a deep understanding of its target base (Myntra, 2019). In fact, Myntra designs a number of cloth lines exclusively based on customers' feedback every year. The customers' feedback also helps Myntra to continuously improve its products, services, packaging, and delivery services. No wonder, social media conversations are proving instrumental in transforming the image of Myntra from merely an online fashion retailer to a lifestyle organization. Going one step further, Hyundai Motors India Limited (HMIL), a wholly owned subsidiary of Hyundai Motors, South Korea, asked netizens to name its small car that was launched in India in 2018. The campaign called "HyundaiNaamKaran" was initiated by the company asking its social media followers to suggest a suitable name for its small hatchback that was known by its codename AH2 before the market launch. The campaign which lasted for 46 days proved extremely popular and witnessed participation from tens of thousands of Hyundai's followers suggesting a range of different names for the car. Out of all names that made their way into the final list, the name "Santro" was overwhelmingly supported by majority of the participants. As a result, Hyundai named its AH2 car Santro and launched in the Indian market on October 23, 2018. Just so you know, Santro was a popular small car model that was first launched in India in 1998. It remained in business for a long period of 14 years before finally Hyundai India pulled a plug on it in 2014 (AutocarIndia, 2015). However, within four years the third-generation Santro made its entry into Indian with improved looks, interiors, equipment, and safety features. The naming campaign ran by Hyundai India for Santro proved extremely effective not only in helping people remind the long association that Santro nameplate has with customers but also build a sales momentum for the new model. Both these objectives were quite effectively met by the Hyundai India with customers flocking to showrooms and Santro registering robust sales of 9009 units in the very first month of its sales in November 2018 (Ram NP, 2018).

Recommendation: Use social media to know deeply about your prospective customers and then use feedback to improve your value proposition. It's not necessary to make dramatic changes as small alterations based on genuine feedback will be good enough to do the trick for the company.

3 Use Social Media Optimally

It's easy to get swayed by the dominance of digital platforms. Many companies nowadays allocate social media a lion's share in their promotional budget, however; putting an excessive focus on social media might blunt its effectiveness. Further, this could also result in a shortage of funds for other critical elements of IMC. As complained by one of the senior brand executives of a cement-producing organization, "It's not logical to allocate 65% of the total promotional budget only to the social media marketing platforms. Yes, we all realize the importance of social media in this day and age but such a heavy reliance hardly makes any sense, especially in our industry where sales promotion, advertisements (print and electronic media), and events make more sense than social media conversations." The observation carries favor with comments made by other executives working in engineering, FMCG, and consumer durable space who all said that dominance of social media in the IMC strategy of the company doesn't bode well for their businesses.

To understand the collective importance of IMC elements, consider the example of the FMCG major Hindustan Unilever Limited (HUL). The Indian subsidiary of the British–Dutch transnational company is the leading brand in the Home-care and Personal-care segments in India. It spends a lot of money on branding and in fact, the company has the biggest promotional budget among all players in the FMCG industry. But the real beauty lies in the manner HUL puts together promotional elements of its IMC strategy in a well-balanced and proportional manner. Given that HUL's target market is deeply entrenched in urban, semi-urban, and rural areas, the company has given proportional weightage to its spending across social media platforms, advertisements, direct selling, and promotional events. Specifically, HUL spends 15–20% of its total promotional budget on social media (effective in covering mostly urban areas) while the rest 80–85% is distributed among other channels in accordance with their importance to the target market (responsible for driving consumption among semi-urban and rural masses). In addition, the allocation of promotion funds also varies according to demographics, and depending on target market characteristics, the company reallocates the budget to reach prospective buyers effectively (Sitapati, 2020). The strategy has largely been successful, thereby helping HUL to stay ahead of rivals in reaching and capturing its target market.

Similar business acumen is displayed by the Indian Automobile Brand Mahindra and Mahindra which has designed its branding strategy with equal focus on print, electronic, and social media channels. Cognizant of the fact that the majority of customers are from middle to upper-middle-income levels residing in villages, towns, and small cities; the homegrown automaker has ensured that specific channels of promotion are used to target particular customer segments. Take, for instance, the Bolero—the best-selling SUV of Mahindra. Thanks to its rugged appeal and solid reliability, Bolero is the preferred choice of customers in areas where road infrastructure is in bad shape, the terrain is challenging, and harsh weather conditions are the norm rather than an exception. The majority of these places are also devoid of good internet connectivity and digital literacy among the people can be best described as Average. Naturally, in such conditions, it doesn't make much sense for Mahindra to promote Bolero through social media and this explains why the majority of Bolero's promotions are done through the regional newspapers, local events, and weekly markets. These mediums of promotions exert considerable influence on the target market as these are part and parcel of their (Philip, 2020). Now we shift focus to another offering of Mahindra, i.e., Alturas G4 SUV. It is the most luxurious SUV in the Mahindra's stable and given the fact that its customer segment belongs to top-level income class, Mahindra exclusively promotes it through social media channels along with selected print and electronic media channels. In social media networks, the major promotion is done through Instagram while in print and electronic media, business newspapers and news channels are preferred choice of the company for promoting Alturas G4 among the target market. Clearly, one can identify a marked difference in Mahindra's promotional strategy which is aligned with the nature of its products, and accordingly, the company allocates fund for its promotional strategy.

Recommendation: We all agree on the importance of social media integration but this shouldn't come at the expense of other elements in the company's branding strategy. Remember, all IMC components such as sales promotion, direct marketing, personal selling, public relations, sponsorships, etc., are important and along with social media, these tools should be given their fair share in the organization's IMC strategy.

4 Align Social Media Integration with Business Goals

'Different organizations have different visions, different business objectives, and therefore, the strategy to integrate social media should also be different. Take, for instance, the case of Zoomcar startup. The self-drive car rental service was founded by David Back and Greg Moran in 2013. Headquartered in Bangalore, the company is currently offering services in 45 different cities in India. The startup is focusing on big cities in the first phase of its market expansion with special focus being given to the metropolitan cities with large corporate offices and business hubs. As the organizations are still going through its establishment phase, it is using social media platforms for creating awareness among the target market, building brand image, testing business ideas, connecting to new customers, rewarding loyal users, and receiving feedback (Sengar, 2020). Aligned with these objectives, the focus of the company's social media conversations is to get more likes, enroll new members, build communities of influencers, incentivize promoters, and award loyal customers. Clearly, Zoomcar has effectively integrated its social media strategy with its business objectives of expansion and new customer acquisitions.

Samsung India, on the other hand, has adopted a starkly contrasted approach for its social media integration strategy. Unlike Zoomcar, the subsidiary of the South Korean consumer-electronics giant is a well-established firm and a household name among Indian buyers. It established its first manufacturing facility in India in 1996 and since then, the company is operating in a range of categories in the consumer durable segment of the Indian market (Samsung Newsroom India, 2018). Samsung is the leading player in Indian Television market while it is among the leading manufacturers of mobile phones, air-conditioners, refrigerators, and washing machines in the country. With an operational legacy of around two and a half decades in India, the company is now focusing on increasing its market share while building brand loyalty among existing customers. Aligned with its business goals is the social media strategy of the company which focuses on driving consumption, boosting market share, facilitating geographic expansions, enhancing real-time engagements, and strengthening brand loyalty among the existing set of customers. This is an entirely different perspective than what we have seen in the case of Zoomcar whose social media strategy is primarily centered on creating brand awareness.

With different objectives, metrics to gauge the success of social media strategy should also be different. While Zoomcar should be concentrating on enhancing the number of likes and followers, Samsung's success metrics must focus on evoking repeat purchases, inducing behavioral changes, or strengthening brand equity.

Recommendation: Depending upon the number of years your organization is in business, brand managers need to customize their social media integration strategy. Management should also define success metrics within the time and budgetary constraints to get desired results. Once you align your social media strategy with your organization's objectives, the likelihood of getting success increases considerably.

5 Select the Right Social Media Platform

Choosing the right social media platform is an important cornerstone of success. Depending on the taste of the target market, you should choose the social media platform to go ahead with. Different demographics have different preferences and to make the right decision, you must do proper research in this regard. Take, for instance, characteristics of generation Z make them more inclined to new-generation platforms such as TikTok, Facebook, or Snapchat. However, generation Y and X consisting mainly of business people and professionals are more likely to patronize LinkedIn and Twitter for their hangout activities. The nature of product/service offerings also influences the choice of social media platforms—it makes sense for a jewelry brand to select Instagram as its preferred platform owing to its emphasis on aesthetics and visual representation. Brand strategists should also focus on the popularity of social media platforms in different countries. Facebook-owned WhatsApp is the most popular messaging app in India while in the Chinese market, WeChat leads social media conversations. Therefore, the choice of selecting a particular social media platform ultimately boils down to the targeted market, their preferences, and the popularity of social media platforms among prospective buyers.

For a new startup or small organization (having a few offerings), selecting a social media platform is not a difficult task. However, for big corporate with multiple product lines, the selection decision could prove to be a tricky one. In such a difficult situation, organizations need to take a cue from Maruti-Suzuki India Limited (MSIL) which is effectively using different social media platforms for connecting with its prospective buyers (Bhatia, 2020). MSIL is the largest carmaker in India and makes cars across categories—hatchback, compact sedan, sedan, and SUV. According to the nature and price point of the product, the company engages with its customers using a mix of social media platforms. For its entry-level budget car, Alto 800, the company makes extensive use of Facebook while for its luxury offerings, i.e., Ciaz Sedan and Vitara Brezza SUV, the engagement with customers is primarily done through Instagram and Twitter. The use of different social media platforms aligned with the price pyramid of its products is helping MSIL to engage effectively with buyers.

Adopting a similar approach is the very popular tea brand Brooke Bond which is owned by Hindustan Unilever Limited (HUL). Brooke Bond is one of the oldest tea brands in the world and offers a range of products across the globe. In the Indian market, its two main product lines include the Red Label and Taj Mahal tea. While the former is targeted at middle-income families, the latter is the premium tea with customers belonging to high-income group. The social media platforms for promoting both brands are different as Facebook is the preferred medium for Red Label advertisement while Instagram is heavily used by the company to promote the Taj Mahal Brand. The strategy seems to be working well as both brands are leading their respective segment in terms of market share and customer satisfaction. Clearly, HUL's decision to select social media platforms in accordance with the price and target market of the brand is yielding rich dividends for the company.

Recommendation: The choice of social media platforms must be made according to the nature of the product proposition. For emphasizing aesthetics, Instagram and Facebook are more suitable choices while Twitter and LinkedIn are more preferred places for informative and serious engagements.

6 Focus on Content

The importance of good quality content couldn't be overemphasized. According to one recent survey conducted by Harvard University, 58% of the respondents said that they will like to follow brands having inspirational content. Despite this, it's ironic to find a lot of companies in the market which have everything in place to pursue their social media strategy except good quality content. Partly responsible for this apathy is a limited understanding of brand executives toward the need to create quality content. These functionaries are unaware that the content quality can significantly impact the engagements of customers. A well-written content with an inspirational message brings vitality to the relationship and uplifts the interaction experience. Quality content also adds to the appeal of the product/service by projecting a positive image in the minds of the target market. In sum, content quality plays an important role in ensuring success not only of the company's social media strategy but also of its overall image and credibility among customers.

The very first step in creating engaging content is to know what kind of content your target audience will likely to follow and engage with. While teenagers prefer pictures and videos, middle-aged people and older adults are more likely to follow blogs, articles, and informational content. The choice of content should also be analyzed from the perspective of the product and services the organization is offering. In case of costly purchase, the focus of the content should be to supply more information (to alleviate the risk factor) although the low-cost and mid-priced items can certainly do away with overloading buyers with information. This explains the long articles and blogs of information that companies in the field of real estate, consumer electronics, and automobile usually publish to help customers make the right decision. The products/services offered by these businesses are high-value items and before purchase, customers usually collect and analyze information to minimize the possibility of making an error in the product selection. The companies too are aware of the risk factor and hence, in order to help customers they supply prospective buyers with as much information as possible. On the contrary, the FMCG sector doesn't feel any such compulsion as the purchase price of products happens to be very low. It means the risk associated with the purchase is low and hence, customers can make do with just regular information about the product.

Another important aspect of the content is its uniqueness and usefulness. Besides, the content should be easily sharable to drive up engagements and sparking conversations. A suitable example in this regard is the graphic released by the World Health Organization (WHO) amidst the ongoing crisis of coronavirus pandemic. The graph shows guidelines to prevent the spread of the highly infectious virus through representational images in a simple yet highly effective manner. Not only the graphic comes across as a useful piece of content but also the small size of the image makes it extremely easy to share. No wonder, the graphic is finding favor with masses and delivering its intended message effectively.

Recommendation: Simple, unique, and sharable content has the potential to engage customers effectively. Prefer the use of images, infographics, and visual content to communicate and help people understand the message in an easy and effective manner.

7 Engage on a Real-Time Basis

Effective social media interactions are two-way communications between the company and its customers. Engaging with consumers on a real-time basis is not a difficult task although most companies fail to seize the opportunity as they remain preoccupied with their own campaigns. However, others are using real-time engagements to their advantage. In order to respond timely and effectively, it's important that the organization must have proper resources—both human and technical. The role of technology is quite crucial here and it's important for the organization to have a proper and well-laid out IT infrastructure in place to respond to customers' queries and their complaints on a real-time basis.

Consider the example of Tata AIG general insurance. The company has a very responsive social media cell and its escalation matrix makes sure that all online complaints must be answered within the four hours of posting. Upon receiving the complaint, the customer care executive makes a follow-up call and an email with a summary of the discussion is sent to the customer within an hour. If the handling of the complaint requires some time, a clear deadline for the resolution of the matter is conveyed to the customer (Aggarwal 2020). No wonder, the company consistently ranks among the top in the consumer grievance handling scale.

Even more important and urgent is to reply to the concern/criticism expressed by celebrities having a large fan following on social media. An organization needs to be more swift and cautious with the response as any mishandling can result in widespread backlash. The largest private sector bank in India, Housing Development Finance Corporation (HDFC) has a dedicated team for handling such situations. Any issue/concern faced by famous personalities is responded on a real-time basis. The team asks the customer to share the issue and contact details through a private message so that the conversation can be taken offline. Further, the company makes sure to put out a response to the complaint on social media so that fans/followers get aware that the organization has taken the note of issue and due procedure is being followed to sort the problem.

Recommendation: Real-time engagement inspires consumers and enhances the credibility of brands. More specifically, it shows that the company is empathetic and willing to listen. Brand Managers must regularly engage with the online community through query handling, suggestions, contests, updates, etc., and responds to their concerns on a real-time basis to make the engagement truly effective.

8 Conclusion

Thanks to digital Technologies, brands today have the ability to connect with customers like never before. Social media can help companies to engage, connect, and inspire consumers to drive the business and powers their growth story. Organizations around the globe are well aware of the importance of social media today. There is consensus among brand executives that social media can prove to be very effective in ensuring the sustainability of their business although how to implement and integrate social media strategies in the overall branding efforts is not very clear. The part of the confusion is due to the lack of knowledge and awareness even among senior branding functionaries who fail to find a fit between social media promotions and the rest of their promotional elements.

In order to help executives conceive and implement social media strategies well, we had interacted with a number of brand executives across business categories. The insights from those conversations were then combined with real-world cases following which recommendations have been made above in the chapter. We sincerely hope that by following these recommendations, an organization can align its social media strategy with its vision, business objective, and customer service mission. This, in turn, will help it to post better financial results riding on the wave of greater market share and higher customer satisfaction.

References

Aggarwal, N. (2020). *Tata AIA life makes 80% transactions self-serviced.* https://cio.economictimes.indiatimes.com/news/business-analytics/tata-aia-life-makes-80-transactions-self-serviced/74644559. Accessed 5 May 2020.

AutocarIndia. (2015). *Hyundai Santro discontinued after 16 years.* https://www.autocarindia.com/car-news/hyundai-santro-discontinued-after-16-years-392752. Accessed 6 May 2020.

Bhatia, S. (2020). *Living the digital dream of car selling.* https://auto.economictimes.indiatimes.com/news/aftermarket/living-the-digital-dream-of-car-selling/75444395. Accessed 1 May 2020.

Myntra. (2019). *Myntra's first digital fashion influencer talent hunt launches soon!* https://blog.myntra.com/myntras-digital-fashion-influencer-talent-hunt/. Accessed 21 March 2020.

Philip, L. (2020). *Auto companies look at Bharat play to boost sentiments during FY21, aims rural market.* https://auto.economictimes.indiatimes.com/news/industry/auto-companies-look-at-bharat-play-to-boost-sentiments-during-fy21-aims-rual-market/75595254. Accessed June 8 2020.

Ram NP, A. (2018). *Hyundai Santro sales higher than Tata Tiago, Maruti Celerio in November 2018.* https://www.rushlane.com/hyundai-santro-nov-2018-sales-12288925.html#:~:text=For%20Nov%202018%2C%20sales%20of,month%20of%20sales%20for%20Santro. Accessed 8 May 2020.

Samsung Newsroom India. (2018). *Samsung Inaugurates World's Largest Mobile Factory in India; Honourable Prime Minister Shri Narendra Modi Flags-off 'Make for the World'.* https://news.samsung.com/in/samsung-inaugurates-worlds-largest-mobile-factory-in-india#:~:text=HC%20Hong%2C%20Chief%20Executive%20Officer,plant%2C%20spread%20over%20129%2C000%20sq. Accessed 12 May 2020.

Sengar, N. S. (2020). *How Zoomcar plans to capture 80 percent market share with its digital-dominated marketing strategy.* http://resource-cms.springernature.com/springer-cms/rest/v1/content/40204/data/v5. Accessed 29 March 2020.

Sitapati, S. (2020). *Hindustan Unilever's advertising approach is a life skill worth acquiring.* https://qz.com/india/1780137/huls-taj-mahal-lux-ad-strategy-in-india-has-lessons-for-all/. Accessed 25 March 2020.

Index

S. Das and S. Gochhait (eds.), *Digital Entertainment*,
https://doi.org/10.1007/978-981-15-9724-4

The manufacturer's authorised representative in the EU is Springer Nature Customer Service Centre GmbH, Europaplatz 3, 69115 Heidelberg, Germany. If you have any concerns regarding our products, please contact ProductSafety@springernature.com

Printed and bound by CPI Group (UK) Ltd, Croydon, CR0 4YY

30/09/2026

02225325-0004